Political Ecology of REDD+ in Indonesia

Indonesia's commitment to reducing land-based greenhouse gas emissions significantly includes the expansion of conservation areas, but these developments are not free of conflicts. This book provides a comprehensive analysis of agrarian conflicts in the context of the implementation of REDD+ (Reducing Emissions from Deforestation and Forest Degradation) and forest carbon offsetting in Indonesia, a country where deforestation is a major issue.

The author analyzes new kinds of transnational agrarian conflicts which have strong implications for global environmental justice in the REDD+ pilot province of Jambi on the island of Sumatra. The chapters cover: the rescaling of the governance of forests; privatization of conservation; and the transnational dimensions of agrarian conflicts and peasants' resistance in the context of REDD+. The book builds on an innovative conceptual approach linking political ecology, politics of scale and theories of power. It fills an important knowledge and research gap by focusing on the socially differentiated impacts of REDD+ and new forest carbon offsetting initiatives in Southeast Asia, providing a multi-scalar perspective.

It is aimed at scholars in the areas of political ecology, human geography, climate change mitigation, forest and natural resource management, as well as environmental justice and agrarian studies.

Jonas I. Hein is a Researcher at the Institute of Geography, Kiel University, Germany, and Associate Researcher at the German Development Institute/ Deutsches Institut für Entwicklungspolitik in Bonn, Germany. He completed his PhD in Human Geography at the University of Goettingen, Germany.

Routledge Studies in Political Ecology

The *Routledge Studies in Political Ecology* series provides a forum for original, innovative and vibrant research surrounding the diverse field of political ecology. This series promotes interdisciplinary scholarly work drawing on a wide range of subject areas such as geography, anthropology, sociology, politics and environmental history. Titles within the series reflect the wealth of research being undertaken within this diverse and exciting field.

Political Ecology of the State
The Basis and the Evolution of Environmental Statehood
Antonio Augusto Rossotto Ioris

Political Ecologies of Meat
Edited by Jody Emel and Harvey Neo

Political Ecology and Tourism
Edited by Sanjay Nepal and Jarkko Saarinen

Environment and Society in Ethiopia
Girma Kebbede

Political Ecology of REDD+ in Indonesia
Agrarian Conflicts and Forest Carbon
Jonas I. Hein

For more information about this series, please visit: www.routledge.com/ Routledge-Studies-in-Political-Ecology/book-series/RSPE

Political Ecology of REDD+ in Indonesia

Agrarian Conflicts and Forest Carbon

Jonas I. Hein

Routledge
Taylor & Francis Group
LONDON AND NEW YORK

First published 2019
by Routledge
2 Park Square, Milton Park, Abingdon, Oxon OX14 4RN

and by Routledge
52 Vanderbilt Avenue, New York, NY 10017

First issued in paperback 2020

Routledge is an imprint of the Taylor & Francis Group, an informa business

British Library Cataloguing-in-Publication Data
A catalogue record for this book is available from the British Library

Library of Congress Cataloging-in-Publication Data
Names: Hein, Jonas I., author.
Title: Political ecology of REDD+ in Indonesia agrarian conflicts and forest carbon / Jonas I. Hein.
Description: Abingdon, Oxon ; New York, NY : Routledge, 2019. | Series: Routledge studies in political economy | Includes bibliographical references and index.
Identifiers: LCCN 2018032992 (print) | LCCN 2018045320 (ebook) | ISBN 9781351066020 (eBook) | ISBN 9781138479319 (hardback) | ISBN 9781351066020 (ebk)
Subjects: LCSH: Deforestation—Indonesia. | Deforestation—Social aspects—Indonesia. | Greenhouse gas mitigation—Indonesia. | Carbon offsetting—Indonesia.
Classification: LCC SD418.3.I5 (ebook) | LCC SD418.3.I5 H45 2019 (print) | DDC 333.75—dc23
LC record available at https://lccn.loc.gov/2018032992

ISBN 13: 978−0−367−58280−7 (pbk)
ISBN 13: 978−1−138−47931−9 (hbk)

Typeset in Bembo
by Apex CoVantage, LLC

Contents

Acknowledgments

Many colleagues and friends helped me to complete this book.

This research would not have been possible without the support of Endriatmo Soetarto, Soeryo Adiwibowo and Rina Mardiana from IPB and Rosyani from UNJA. I am really grateful for your advice, for the opportunity to benefit from your great networks, for engaging with RISTEK and for your inspiration. Moreover, I would like to thank Heiko Faust, Fariborz Zelli and Christoph Dittrich. I am really thankful for your support, for engaged discussions and your supportive pragmatism.

At CAU Kiel, I would like to thank Silja Klepp for constructive feedback on chapters, giving me the time to finish this book and for the generous support of her chair for copy editing. Moreover, I would like to thank Sarah Hartwig, Mytree Delfs and Lasse Nissen for their great support.

At my former workplace, the German Development Institute, I would like to thank especially Carmen Richerzhagen. Thank you, Carmen, for your critical and encouraging feedback, commenting on draft chapters, manuscripts, presentations and proposals. Furthermore, I would like to express my gratitude to Ines Dombrowsky. Thank you, Ines, for your trust and for providing me the freedom to develop my own ideas. Special thanks to Jean Carlo Rodriguez. I really benefitted from our critical discussions, extended coffee breaks, your knowledge and your feedback.

A very huge "thank you" to all student and research assistants from IPB, UNJA, University of Göttingen and German Development Institute, especially to Arie, Syakir, Niklas Rehkopp and Peter Hermes. Thanks Arie and Syakir for your company, for traveling with me, for your energy and your efforts. It was a great time. Thanks Niklas and Peter for your patience during endless phone calls and for preparing most of my maps for papers, conference presentations and for this book.

Moreover, I am thankful to the project coordinators Wolfram Lorenz in Bogor, Bambang Irawan in Jambi and Annette Koehler-Rahm in Bonn. You did a great job!

Thank you Yvonne Kunz, Barbara Beckert, Stefanie Steinbach and Michael Miessner for your constant support, intense discussions and for finding answers. I wish you all the best!

On a personal note I would like to thank my friends and family. First of all, to my dearest Marie and to my son Ben; your love gave me the force for staying on track during this long way. I would also like to thank my friends Thorsten, Björn and Alex for all your personal support. I am especially grateful to my family, to Wolfgang, Jamin and Brigitte for all your care and trust.

Last but not least I would like to thank all the people that I interviewed, the village communities of Air Hitam Laut, Bungku, Tanjung Lebar, Seponjen and Kampung Laut, and SPI and ZSL for supporting my field stays. Terimas Kasih Banyak!!!

Figures and tables

Figures

Tables

1 Introduction

When traveling through the dense oil palm plantations around the village of Bungku to the "forest of hope" (Harapan Rainforest, Indonesian: *Hutan Harapan*)[1] in southeastern Jambi on the Indonesian island of Sumatra in 2012, the people I met were troubled and rebellious. For almost 30 years, they have been fighting against the violent appropriation of community land by a transnational oil palm company. More recently, a new conflict has emerged to affect their lives – after the conservation company PT Restorasi Ekosistem Indonesia (REKI), founded by a transnational NGO (nongovernmental organization) consortium and financed by the German and Danish governments, had received a conservation concession from the Ministry of Forestry.[2] The new concession further reduces the land available for the peasants and indigenous communities of Bungku and neighboring villages. The village of Bungku, which the inhabitants, after years of conflict, describe as the "village of 1001 problems", is now effectively a village without land, stuck between the dark green of corporate oil palm plantations and PT REKI's new Harapan Rainforest conservancy.

Yet, there are signs of resistance along the road. Signs and boundary stones placed by villagers demarcate land claims. Banners and wooden portals welcome visitors to the territory of the indigenous Batin Sembilan group and to officially inexistent peasant settlements in the "forest of hope". When talking to peasants, you rapidly realize that it is not their "forest of hope" but a forest charged with stories of conflict and resistance. Some villagers allege that the Prince of Wales, who visited the area in 2008, now owns the forest. Others explain that they cannot use their rice swiddens anymore since the government declared the forest "as the lung of the earth". Some of those living in the Harapan Rainforest tell you that "the rich countries bought the oxygen in the forest", while others complain that their "home is not the carbon toilet for the rich countries", asking why Germany and Denmark are not reducing their greenhouse gas emissions at home instead of protecting forests in Indonesia.

Many peasants explained that their presence within the conservation concession was not accepted by the conservation company. They told me about clashes with the army, "black" private security agents and the riot police, about houses that had been burned supposedly by private security and forest police, and about brave women organizing sit-ins in front of bulldozers and heavily armed policemen. When entering the Harapan Rainforest, the tensions were visible. Visitors had to pass security checkpoints, the main camp was fortified, and the riot police, "BRIMOB", had established a temporary base close to an ancient airfield within the conservation concession. PT REKI's staff found themselves in an emergency situation after staff had been kidnapped by peasant activists to use as bargaining chips in negotiations for the release of peasants who were arrested by private security and the police. Both parties accused each other of being responsible for the violent escalation of the conflict and the anarchic conditions.

Protest and resistance in and against the Harapan Rainforest conservancy took many different forms and involved various actors. Some peasants received support from peasant organizations and organized, for instance, a march to the Ministry of Forestry in Jakarta. Village governments ignored the borders of the concession and issued land titles for plots located within it, while others rather engaged in hidden activities, such as logging. However, not all peasants were engaged in protest against PT REKI. Some members of the indigenous Batin Sembilan consider the conservation company to be an important ally against the expansion of agro-industrial oil palm estates. They explained that their forest gardens and swiddens had been destroyed by oil palm companies. They hope that the conservation project will help them to reestablish their traditional livelihoods.

This book focuses on these new stories from the Harapan Rainforest and many others from other places on Sumatra, such as the Berbak Carbon Initiative. The stories center on the socio-spatial consequences of a recent invention by economists and conservationists in the context of international climate policy: the REDD+ (Reducing Emissions from Deforestation and Forest Degradation) mechanism. They stand for a new kind of conservation conflict linking different places and actors, such as developed countries and corporate actors interested in carbon offsetting, conservationists interested in financing protected areas by selling carbon credits, and peasant farmers and indigenous groups in the Global South interested in maintaining or expanding access to agricultural land. The conflicts over Sumatra's remaining forests are symptomatic for the contradictions and ambiguities caused by attempts to maintain the current fossil-fuel based accumulation regime and by attempts to export climate change mitigation to supposedly low-cost locations in the Global South.

In Indonesia, a large highly biodiverse archipelago with the third largest extent of tropical forest cover, the situation is notably conflictive. On the one hand, Indonesia claims to be a global leader in REDD+; on the other hand, local realities such as land tenure conflicts and the rampant forest fires of 2015 stand in sharp contrast to Indonesia's announcements concerning forest governance reforms (Finlayson, 2014; Hein et al., 2016: 380; Toumbourou, 2015). In particular, Indonesia's last remaining frontier areas are heavily contested spaces that are still witnessing violent conflicts about access and control of forest land (Hein et al., 2016; Tsing, 2005). Caused by historically contingent structural inequality, land conflicts "became chronic" (Rachman, 2013: 3) in Indonesia.

Colonial and post-colonial governments appropriated vast forest areas for resource exploitation and conservation. Today, indigenous communities and peasant organizations consider the occupation of corporate and state-owned plantation estates and conservation areas as a legitimate response to the dispossessions of the colonial and post-colonial state (Hein et al., 2016; Lukas, 2014; Peluso et al., 2008).

Indonesia has 252 million inhabitants and the fourth-largest population in the world, the third-largest extent of tropical forest cover and one of the highest rates of land-based greenhouse gas emissions, mainly caused by forest fires and deforestation. In this context, the Indonesian government, supported by Norway's International Climate and Forest Initiative (NICFI), selected a number of Indonesian provinces for piloting REDD+ policies. The province of Jambi, located on the island of Sumatra, became one of these provinces in 2013.

This book shows that REDD+ policies are seldom conflict free, especially when local communities have to cope with the simultaneous expansion of agro-industrial estates. The book focuses on two different but interrelated forms of peasant resistance and agrarian conflict. The first refers to land occupations that occurred before REDD+ implementation and that were organized by village governments, indigenous leaders and peasants, and facilitated by sub-national branches of the state. The second refers to resistance and attempts to defend land rights during the implementation of REDD+. Taking a political ecology perspective, the aim is to unravel the causes and the scalar dimension of land conflicts and peasant resistance by asking questions, such as: Which actors are involved in land tenure conflicts in REDD+ target areas? What are the historical root causes of conflictive property rights? Which role does power play? What are the explanations for conflict? What are the climate justice implications of transnational forest conservation initiatives such as REDD+?

Introducing the politics of REDD+ and peasant resistance

When the idea of reducing emissions from deforestation entered the UN climate negotiations in 2005, this was also the result of lobbying by an interesting transnational actor coalition. This transnational coalition consisted of a number of tropical forest countries represented by the Coalition of Rainforest Nations, large environmental NGOs and a number of large transnational companies interested in the cost-efficient offsetting of greenhouse gas emissions (e.g. Hein and Garrelts, 2014; Jodoin and Mason-Case, 2016; Stephan et al., 2014). Influential reports such as the *Stern Review* on the "Economics of Climate Change" (Stern, 2007) and Johan Eliasch's (2008) review "Climate Change: Financing Global Forests" identified forest conservation as the most cost-efficient way to mitigate climate change. In addition, the third assessment report of the Intergovernmental Panel on Climate Change (IPCC) published in 2007 pointed out that 17 percent of global greenhouse gas emissions are caused by deforestation and forest degradation. These key publications supported the growing momentum for the inclusion of forest conservation in the United Nations Framework Convention on Climate Change (UNFCCC) agenda (Jodoin and Mason-Case, 2016).

The basic idea behind REDD+ was taken from dominant market-oriented conservation concepts, such as payments for ecosystem services and carbon trade. Payments to forest owners or tropical forest countries channeled through carbon markets, global funds or national payment for ecosystem services schemes would halt deforestation almost automatically. Supporting forest conservation in the Global South was framed as a "win-win" solution that would contribute to biodiversity conservation, rural development and climate change mitigation (Angelsen et al., 2012; Pagiola, 2011; Virgilio et al., 2010; Visseren-Hamakers et al., 2012). Thus, the idea rapidly gained political traction and also received support from previously skeptical actors, such as the European Union.

In 2007, REDD+ became part of the Bali Roadmap towards a post-2012 binding climate agreement. In contrast to previous market mechanisms and climate finance mechanisms, such as the Clean Development Mechanisms and the Adaptation Fund of the Kyoto Protocol, REDD+ is not governed by a central management or financing body. No final agreement has yet been made on how REDD+ will be financed, and whether emission reductions from REDD+ activities should be considered as voluntary contributions from developing countries supported by the developed countries, or whether REDD+ should be eligible for offsetting (Angelsen et al., 2014; Horstmann and Hein, 2017; Jodoin and Mason-Case, 2016). In fact, today's REDD+ governance is highly fragmented (Zelli et al., 2014). This fragmentation is also reflected at the national and local scales (ibid.). The UNFCCC is only one of many institutions governing REDD+. Most REDD+ country programs and local projects are funded by bilateral donors, such as Norway, the United Kingdom and Germany, by multilateral donor agencies such as the Forest Carbon Partnership Facility (FCPF) administered by the World Bank, and through voluntary carbon markets. All these actors have developed their own specific ideas on the recognition of community rights, on how to reduce deforestation, on rules for financing forest conservation, and on achieving supposed win-win outcomes. Some projects aim to produce emission reduction certificates for voluntary carbon markets (as the Berbak Carbon Initiative discussed in this book). Indonesia, Colombia, Guyana and Brazil, for instance, have negotiated result-based payment agreements with donors, meaning that they receive payments if they successfully reduce deforestation. Other donor-financed REDD+ projects aim to "improve" national forest governance or to make countries "ready" for carbon markets. In some cases, conservation NGOs and national conservation authorities strategically link their existing conservation efforts to REDD+ to gain access to donor finance.

Consequently, REDD+ can be many different things at different scales and for different actor coalitions and their respective storylines and discourses. In particular, because of its fuzziness, REDD+ still is and was very popular as an idea. For some actors, REDD+ is a cost-efficient offsetting mechanism, for others rather a development aid mechanism to support transformation towards rural low carbon economies, for environmental NGOs and often underfunded national conservation authorities it is rather a mechanism to finance protected areas. Indigenous communities across the globe use REDD+ to access development aid and to argue for land rights, while peasant associations consider REDD+ as enabling land grabs for the purpose of forest carbon offsetting (La Via Campesina, 2015). For those that Hiraldo and Tanner (2011) describe as "institutionalists", REDD+ is an attempt to establish a global scale of "good" forest and land tenure governance to support climate change mitigation, biodiversity conservation, sustainable development, and the rights of local and indigenous communities (Gupta, 2012: 622; Hiraldo and Tanner, 2011: 46).

From a global environmental and climate justice perspective, REDD+ raises a number of additional concerns. REDD+ focuses on developing countries and especially on the peripheral forest margins. Forest conservation and the transfer of financial resources could help to promote more sustainable land use, but it also transfers responsibility for reducing greenhouse gas emissions to those who have emitted much less than citizens in the Global North. Furthermore, as a recent study of Irfany and Klasen (2015) shows, Indonesian citizens living in urban areas emit twice as much as those living in rural areas. From the perspective of an urban citizen either from Indonesia or from the Global North, REDD+ is attractive because it transfers obligations either to the global periphery or to the national periphery, avoiding hard emission cuts that would affect their own lifestyles. The expansion of protected areas and reforestation efforts rather affect the rural population. Consequently, the factual or putative lower-opportunity costs of forest conservation are not politically neutral.

Indigenous groups and peasants in many parts of the world criticize the "global gaze" (Fogel, 2004) of the REDD+ mechanism and the framing of forests as empty carbon stocks, highlighting that many indigenous groups and peasants live within and close to forests and have maintained the carbon storage capacity of forests for generations. REDD+ and green enclosures affect actors differently, reflecting power imbalances at the forest margins but also between the North and South, and between urban centers and rural areas in the South (Eilenberg, 2015; Hein et al., 2018a; Kosoy and Corbera, 2010; Lohmann, 2008; McAfee, 2012b). Following this argument, REDD+ can be considered as a mechanism that stabilizes the current fossil fuel-based accumulation regime (Hein et al., 2018a) and the "imperial mode of living" (Brand and Wissen, 2012, 2017) characterized by high-emission lifestyles and consumerism in global centers. The imperial mode of living is based on a disproportionate claim to global sinks, including to the world's tropical forests, to offset the externalities of high emission lifestyles (ibid.).

In this context, where land and nature are becoming increasingly valuable as carbon sinks, as ecosystem service providers and as fertile grounds for the expanding agro-industry, "[. . .] the basic questions of the agrarian political economy are as relevant as ever: Who owns what? Who does what? Who gets what?" (Fairhead et al., 2012: 241). Land is at the heart of the agrarian question.

It is one of the most important means of production in the agrarian political economy. Providing and legitimizing access to land is an important source of political authority (Lund, 2016). In many target countries of REDD+, it is exactly this political and economic resource that is highly contested (Eilenberg, 2015; Hein et al., 2016, Kunz et al., 2016; Larson et al., 2013; To et al., 2017). Two important reasons for this contestation are as follows. First, especially in Indonesia but also in South American countries like Colombia, the central state has appropriated large parts of the countries' forests, often ignoring the presence of indigenous and customary communities. The formation of the state forest estate (*kawasan hutan*) in Indonesia and the forest reserve (*reserva forestal*) in Colombia were state territorialization projects aimed at claiming land and stabilizing the national territory by allocating land in frontier areas to citizens and companies, challenging pre-existing authority and property relations (Hein et al., 2016; Ortiz, 1984: 210). In both countries, this process was notably violent and was characterized by multiple periods of primitive accumulation (Del Cairo et al., 2014; Escobar, 2003; Gómez et al., 2015; Hein et al., 2016; Peluso, 1995; Peluso et al., 2008). In Indonesia, after the fall of former Indonesian president Suharto at the end of the 1990s, power constellations changed the scalar structure of the state, and this state territorialization project came under serious pressure driven by protests from customary communities, peasant movements and local governments (Barr et al., 2006; Hein et al., 2016; Peluso et al., 2008). Rescaling widened the agency of local political authorities, creating the momentum to exercise de facto control over parts of the state forest territory. Second, land is often very unequally distributed. In the Indonesian village of Bungku, for instance, located at the margins of the Harapan Rainforest, two-thirds of the village land has been allocated to oil palm, timber and conservation companies, leaving only a few hectares available for peasant farming (Polsek Sungai Bahar, 2011; Zainuddin, 2013).

Protest and resistance from peasants and indigenous communities against REDD+ and other conservation initiatives on the island of Sumatra and beyond revolve mainly around the basic questions of the agrarian political economy, as raised by Fairhead and colleagues (2012), and in particular around access to land, land rights and land-use restrictions. However, a particularity of peasant resistance against REDD+ on Sumatra is the explicit reference to global climate justice issues, as illustrated by the slogan introduced above: "Our forest is not the carbon toilet of the rich countries". The basic idea of offsetting emissions at low-cost locations not only links emitters in the North with project implementers and land users in the South, but offsetting also links local struggles on access and control of forest and land resources to transnational activists' networks that provide peasants with the opportunity to resist the land claims of private or public conservation agencies (Chatterton et al., 2013; Hein and Faust, 2014; Hein et al., 2016).

Peasant resistance and indigenous struggle for recognition and rights have been widely discussed by scholars, in particular by James C. Scott (1985). Scott analyzed hidden peasant resistance strategies, arguing that hidden and everyday forms of peasant resistance do not openly challenge hegemony. Rather, they occur silently as the hidden encroachment strategies of peasants entering the forest of hope. But when the political context permits, for example after the fall of Suharto, hidden resistance can turn into open forms of resistance, such as the public invasion of property (Peluso, Afif and Rachman, 2008; Scott, 1989: 5; Turner and Caouette, 2009: 11). The cases discussed in this book fall between the two categories. They include open and organized forms of land occupation, the formation of counter territories and open protest at climate summits, but also hidden encroachment and sabotage. Furthermore, they include resistance against historically rooted and contingent structural inequality and unequal land distribution prior to conservation interventions, and resistance and conflict in the context of ongoing REDD+ and conservation interventions. Whereas peasant resistance can be considered as class struggle (e.g., Scott, 1989), indigenous groups' struggle mainly strives for the acknowledgement of full citizenship rights (especially in Indonesia) and the obtaining of specific minority rights in compensation for being historically disadvantaged, exploited and marginalized by European colonizers and post-colonial governments.

Recently, the social struggles of peasants and of indigenous groups have become more and more transnational. Transnational peasant organizations such as La Via Campesina and the Asian Peasant Coalition emerged out of protest against market liberalism in the 1990s (Borras, 2008). Transnational peasant protest has become a common feature of international trade conferences and, more recently, of climate change conferences. La Via Campesina, as the largest peasant organization, and large indigenous rights organizations, such as AMAN (Alliansi Masyarakat Adat Nusantara) from Indonesia and COICA (Coordinadora de las Organizaciones Indígenas de la Cuenca Amazónica) from the Amazon basin countries, have become key actors of the global climate justice movement (Claeys and Delgado-Pugley, 2017). COICA and AMAN use the REDD+ momentum to lobby for indigenous rights. They initiated, for example, the "no rights, no REDD+" campaign, they suggested a number of indigenous approaches towards REDD+ (e.g., *REDD+ Indigena* of COICA), and strategically use their supposed role as forest stewards to strengthen indigenous rights to land. Peasant associations such as La Via Campesina, in contrast, strictly oppose REDD+ but they also use the additional agency provided by the REDD+ momentum to lobby for peasant rights and food sovereignty and against offsetting and carbon markets (Buckley, 2018; Claeys and Delgado-Pugley, 2017; La Via Campesina, 2015). Both peasant and indigenous movements have strategically used the additional agency provided by global climate governance for transnational campaigns. Both have heavily criticized market-based and state-led strategies to mitigate climate change (Claeys and Delgado-Pugley, 2017: 325).

However, peasants and indigenous communities organized in social movements are not the only type of resistance against REDD+ occurring in Sumatra's forests. The ongoing conflicts also take place within the state. There is competition between the conservation and development-oriented apparatuses of the state, reflecting the different concerns of competing groups within society but also those of supra-national planning institutions (Hein et al., 2018a). In some cases, peasants and indigenous groups have formed alliances with local state authorities and local NGOs in order to regain access to land officially controlled by other state actors and private companies. Especially after the political turmoil of the late 1990s and early 2000s, district heads (Bupati), village heads (Kepala Desa) and customary leaders took advantage of the confusion, interpreting the decentralization reforms to their advantage and asserting far-reaching administrative authority over forests (Barr et al., 2006). In contrast, local conservation authorities used the REDD+ momentum after the climate conference in Bali in 2007 to form alliances with public donors and transnational conservation organizations to gain support for their chronically under-funded protected area system (Hein et al., 2018a).

This book argues that conflict in the context of forest carbon offsetting and REDD+ is a new empirical phenomenon calling for an innovative conceptual framework. The concepts of power, territory and scale provide important explanations for contemporary agrarian conflicts in the context of conservation interventions. REDD+ and broader processes of state transformation have induced significant rescaling processes, transforming the governance of land and forests and producing new forms of territoriality (Castree, 2008; Cohen and McCarthy, 2014; Peluso and Lund, 2011; Reed and Bruyneel, 2010). Investigating the politics of scales sheds important light on "[. . .] who will have access to what kind of nature [. . .]" (Swyngedouw, 2010). Political scales can be considered as spatial delimitations of power relations (Meadowcroft, 2002). Investigating territorialization processes helps to unravel discursive strategies, the construction of social identities, conservation logics and power relations inscribed in landscapes (Peluso and Lund, 2011). Power relations in this vein are important constitutive elements of the social production of space, scale and territory.

REDD+ and access to land are both issues that are negotiated and regulated on different scales, for example at the village scale by the village head, by the district head or (in the case of forest land) by the Ministry of Forestry in Jakarta, or (in the case of REDD+) at UNFCCC conferences and in the headquarters of donor agencies. Consequently, this books builds on multi-sited qualitative research inspired by "deterritorialized" (Merry, 2000: 130) and "multi-sited ethnography" (Marcus, 1995: 80). The main sites of the empirical investigation of this book are villages located at the margins of and within the Harapan Rainforest and the Berbak Carbon Initiative on the island of Sumatra, Indonesia (Figure 2.1). Both projects are located in the province of Jambi. The Harapan Rainforest is located in the southeastern part of Jambi's lowlands, whereas the Berbak Carbon Initiative is located at the coast of the South China Sea. In both projects, transnational NGOs and donors play an important role. In addition, the book is based on qualitative research conducted along the lines of different networks of interaction linked to project implementation and resistance across field sites, namely the provincial capital of Jambi, the national capital Jakarta, the city of Bogor as an important hub for academia and conservation NGOs, UNFCCC climate change conferences and the headquarters of donor agencies in Germany.

A guide through the book

The outline of this book is as follows. Chapter 2 provides an analytical and methodological framework for investigating new agrarian conflicts. Conflict in the context of forest carbon offsetting and REDD+ is a new empirical phenomenon calling for an innovative conceptual framework. Based on the politics of scale literature, political ecology and theories of power, the book aims to advance our theoretical understanding of the new agrarian conflicts emerging in the context of REDD+. It is argued that the ability to alter political scales and consequently the ability to access land and define "nature" is linked to questions of power.

Chapter 3 investigates the politics of the scale of forest and land tenure governance as the dynamic context of REDD+ implementation and agrarian conflicts in Indonesia. The chapter shows how different political regimes, recentralization and decentralization have facilitated the construction of sometimes competing and contradictory political scales governing access to land and forests. Recently, village governments supported by customary authorities have been able to establish village scales of land and forest tenure regulation and to expand village and customary territories, overlapping with the state forest territory and with national scales of regulations. Allied with customary authorities, village governments control land, challenge the integrity of the state forest territory and are able to resist centralized control of forest and land resources.

Chapter 4 focuses on the privatization and transnationalization of conservation in Indonesia. Private companies and donor governments from the Global North consider forest conservation in the South a cost-efficient option to mitigate climate change. Environmental organizations and public conservation agencies in the South are mainly looking for new options to finance tropical forest conservation. The Indonesian government has initiated a number of governance reforms and established REDD+ pilot provinces in order to be "ready" for foreign investments in carbon conservation.

Chapter 5 focuses on the transnational dimensions of agrarian conflicts and peasant resistance. It investigates the different territorialization strategies of peasants, indigenous groups, conservation NGOs and apparatuses of the state. It shows successful examples of peasant resistance that challenge the commodification of forest carbon, and disentangles conflicts between development- and conservation-oriented apparatuses of the state.

In Chapter 6, the book concludes that apparently local land conflicts about access and control of forests and agricultural land become transnationalized in the context of REDD+ and forest carbon offsetting. REDD+ pilot projects in Jambi financed by private and public donors changed the dialectical relationships between structure and agency. They reduced the ability to access land for some actors, provided additional opportunities for others, and provided entry points for the transnational resistance campaigns of peasant movements and climate justice organizations. Successful peasant resistance, as the struggles discussed in this book show, has relied on scale jumping and on transnational support networks. Peasant movements and indigenous peoples have gained agency in the course of recent attempts to establish a global scale of forest governance. Finally, transnational conservation initiatives and market-based conservation instruments such as REDD+ are not acting in a social and political vacuum. Understanding the specific history of landscapes is of key importance for understanding land conflicts triggered by conservation interventions.

Notes

1 The official names are Harapan Rainforest and *Hutan Harapan*. In this book Harapan Rainforest and "forest of hope" as an English translation will be used interchangeably.
2 In 2015, the former Ministry of Forestry (*Kementerian Kehutanan*) and the Ministry for the Environment (*Kementerian Lingkungan Hidup*) were merged. When I refer to the Ministry of Forestry (MOF), I always refer to the institution prior to the merger, when I refer to the Ministry of Environment and Forestry (*Kementerian Kementerian Lingkungan Hidup dan Kehutanan Republik Indonesia*), I refer to the ministry in its current form.

2 Conceptual, theoretical and methodological underpinning for a political ecology of transnational agrarian conflicts

Conflict in the context of forest carbon offsetting and REDD+ is a new empirical phenomenon calling for an innovative conceptual framework and methodological approach. This book aims to advance our theoretical understanding of new agrarian conflicts emerging in the context of REDD+. Building on Edward Soja (1989: 60), I suggest that any analysis of specific geographies of capitalism is necessarily an "eclectic exercise". Especially research on new empirical phenomena that links actors across scales and spaces in a very specific and novel way – such as REDD+ – has to be flexible and creative and should avoid too rigid categorical thinking (ibid. 73). REDD+ and agrarian conflicts are negotiated and regulated on various scales, e.g. at the village scale by the village head, by the district head or – in the case of forest land – by the Ministry of Forestry in Jakarta, or – in the case of REDD+ – at UNFCCC conferences and in the headquarters of donor agencies. Consequently, the empirical research for this book was "multi-sited" and inspired by "deterritorialized" (Merry, 2000: 130) and "multi-sited ethnography" (Marcus, 1995: 80).

This conceptual and methodological chapter outlines the main elements of a political ecology of REDD+, combining critical social theory, Marxist and post-modern geography, and legal anthropology, and provides information on the multi-sited qualitative approach used for the empirical research. It introduces concepts of power, the state, space, territory, scale and property which help to investigate agrarian conflicts in the context of conservation interventions. I argue that REDD+ and broader processes of state transformation have induced significant rescaling processes, transforming the governance of land and forests and producing new forms of territoriality (Castree, 2008; Cohen and McCarthy, 2014; Reed and Bruyneel, 2010; Peluso and Lund, 2011). Investigating the politics of scales explains in important ways "[. . .] who will have access to what kind of nature [. . .]" (Swyngedouw, 2010: 12). Political scales can be considered as spatial delimitations of power relations (Meadowcroft, 2002). Investigating territorialization processes helps to unravel discursive strategies, the construction of social identities, conservation logics and power relations inscribed in landscapes (Peluso and Lund, 2011). Power relations in this vein are important constitutive elements of the social production of space, scale and territory. Thus they are important explanatory factors for the differing abilities of actors to access land and property (Corbera and Brown, 2010; Koch et al., 2008; Nuijten, 2005; Ribot and Peluso, 2003; Rodriguez de Francisco and Boelens, 2014; Wynberg and Hauck, 2014;).

Political ecology

Political ecology explores the production of different "politicized environments" (Bryant, 1998) in the context of specific power constellations. Political ecologists built on a wide range of schools of thought coming from very different epistemological perspectives (e.g. Marxist, hermeneutics/social constructivism, post-structuralism) loosely drawn together by the objective of investigating the interrelations between political economy, ecological processes and power relations. Fundamental for Marxist political ecology is the basic assumption that the appropriation and transformation of the biophysical environment through labor produces a "second nature" (Lefebvre, 1976b: 15) or "social nature" (Castree, 2001).

In German-speaking countries the concept of societal relationships with nature (*Gesellschaftliche Naturverhältnisse*) has been quite influential for scholars of political ecology (Görg, 2011: 416; Köhler, 2008: 210; Pichler, 2014: 19). To some extent the concept of societal relationships goes beyond Anglo-Saxon political ecology, stressing the importance of institutions and different state agencies (Görg, 1999, 2011). The concept explicitly considers that "[. . .] conflicts over societal relationships with nature are closely interlinked with spatio-institutional transformations of the state" (Brand et al., 2011: 150). In line with the societal relationships of nature concepts, this study builds on a dual understanding of nature (Görg, 2011: 416). First, a material dimension of nature refers to economic and technical forms of appropriation (ibid.). Second, a symbolic dimension refers to nature as a cultural construction (ibid.). These two dimensions are not binary opposites, they are intrinsically linked to each other.

Nature is produced by society through practices, through linguistic and scientific meanings, but nature still has a biophysical and material basis with inherent physical processes (Escobar, 1999: 3; Görg, 2011: 417). In more illustrative terms, the social practices of smallholders, conservationists and logging companies transform the meanings of nature but at the same time they transform the biophysical materiality of nature. The biophysical materiality of nature is rather shaped by social practices and discourses than determined by its materiality. Thus, the very same nature, or more specifically a forest, can be experienced differently "[. . .] according to one's social position" (Escobar, 1999: 5). For a logging company a forest is primarily a source of timber. The semi-nomadic Batin Sembilan in the Harapan Rainforest might conceive the same forest as a fruit garden, whilst a frontier migrant might consider the forest as empty space for agricultural expansion. In contrast, a REDD+ project developer might conceive the very same forest as carbon storage and as a "production site" for forest carbon credits.

Linking social-spatial theory with conservation territories and property relations

A socio-spatial perspective on REDD+ provides a number of highly relevant insights, shedding light on the historical and social aspects of spatiality (Soja, 1996). It helps to investigate the formation of territorial units such as protected areas and the social construction of nested scales of governance, to question the crisis narratives that are used to legitimize green enclosures, and to think about how REDD+ – as a market-based conservation mechanism – reproduces existing patterns of uneven spatial development. Jessop and colleagues argue that "[. . .] socio-spatial theory is most powerful when it a) refers to historically specific geographies of social relations, and b) explores contextual and historical specific variations in the structural coupling, strategic coordination, and forms of interconnection among the different dimensions of the latter" (Jessop et al., 2008: 392).

The social production of conservation spaces

Protected areas are social spaces that Zimmerer (2000: 358) describes as "nature-society hybrids" containing second nature. "Environmental conservation makes both space and spatial scale" argues David M. Hughes (2005: 157). The implementation of protected areas redefines space, implies certain rules and designates boundaries, thus conservation implies the "making of territory" (Zimmerer, 2000: 358). In this book, space is considered as a social product. Based on post-modern and Marxist critical theorists, such as Henry Lefebvre and Edward W. Soja, I consider space and spatial units such as forested landscapes as socially produced (Lefebvre, 1976a; Lefebvre, 1991; Soja, 1980). Soja (1989: 79–80) explains that "space in itself may be primordially given, but the organization and meaning of space is a product of social translation, transformation and experience". The social space or created space, according to Soja (1980: 2010), is socially produced through transforming "[. . .] the given conditions inherent in life-on-earth". Lefebvre argues in the same direction. He claims that "space has been shaped and molded from historical and natural elements, but this has been a political process" (Lefebvre, 1976a: 31). For Zimmerer (2000) conservation areas "belong(s) fully to the production of nature and space in the transition to a late capitalist modernity" (ibid. 360). Hence, socially produced space reflects modes of production, political organization and ideology (Lefebvre, 1976a: 31; Soja, 1980: 210).

Protected areas, as an environmentalist spatial practice (Hughes, 2005: 157), contain space. By establishing new conservation areas and by implementing different zones for conservation and for "sustainable" resource use, conservationists fix non-human species and the human population in space. Protected areas are the spatial manifestation of what Garland (2008: 61) and Brockington and Scholfield (2011: 83) describe as the conservationists' mode of production which "[. . .] incorporates wildlife into the capitalist system". Conservationists not only raise money for new protected areas, they also use space to produce a number of commodities such as scenic beauty, spectacular images of endangered flagship species, sites for ecotourism and – more recently – carbon credits. Garland (2008: 62) argues that "[. . .] as is the case for other kinds of natural resource exploitation, the relations of production involved in wildlife conservation depend greatly on the control over productive assets in question [. . .]", which in most cases is space.

In this sense, the social production of conservation spaces are dialectical processes. Space, political organization and relations of production mutually depend on each other (Soja, 1980: 211). Conservation spaces such as the Berbak Carbon Initiative are the outcome of social relations. And, at the same time, by imposing certain rules they structure and mediate social relations.

Scale

The concept of scale expands the socio-spatial theory of Lefebvre and Soja since it locates social practices not as fixed within space but rather as within dynamic socially produced scales (Brenner, 1998: 459, Wissen, 2008: 19). The starting point is the premise that scale is a socially produced hierarchical spatial element (Jessop et al., 2008: 393; Towers, 2000: 26). In contrast to the international relations and multi-level governance literature, I do not take the multi-scalar and hierarchical organization of the state for granted (Bulkeley, 2005: 876). Hierarchical scalar structures of the state including territorial units and their institutionalized forms and levels of representation in Indonesia (e.g. provincial government, district government and village government) are the outcome of societal struggle and political negotiation (Brenner, 1998; Hein et al., 2018a; Houdret et al., 2014; Swyngedouw, 2010). They demarcate arenas of socio-political struggle and regulation. "Scales of regulation" (Towers, 2000), for instance, refer to spatial entities such as nested jurisdictions with a specific regulatory order (Hein et al., 2016). Actors might produce additional scales or shift political struggles to a specific scale of regulation or political forum to pursue their interests (Hein et al., 2018a; Smith, 2008: 232; Towers, 2000; von Benda-Beckmann, 1981).

Scale theorists and political ecologists share many theoretical and empirical concerns and have recently been increasingly engaged in an intense dialogue (Neumann, 2009: 398–400). Early political ecologists such as Blaikie and Brookfield (1987: 79) already highlighted the interconnectedness "[. . .] of political economic relationships at the local, regional and international scales which determine the actions of land-user [. . .]". Authors such as Erik Swyngedouw (2010), Karl Zimmerer (2000, 2006) and Alina Brad (2016), as well as Ulrich Brand and colleagues (Brand et al., 2011), have bridged both strands of literature drawing on the concept of scale to analyze the production and the societal relationships of specific natures (Neumann, 2009: 402).

Three aspects of the current academic discussion within human geography, political science and anthropology are relevant for understanding the different agrarian conflicts in the context of REDD+ in Indonesia. First, scale theorists focus not on scale as such but seek to explain social processes that lead to the production of scale and the transformation of existing scalar arrangements. Neil Brenner, for instance, has conceptualized the role of the state in the production of scale (Brenner, 1997, 1998). Drawing on Lefebvre, he argues "[. . .] that the territorial state has played a crucial role in constructing a worldwide 'second nature' of socio-spatial configurations organized on multiple, overlapping spatial scales" (Brenner, 1997: 149). He further argues that globalization, as ongoing rescaling processes of the world economy, is associated with state interventions that support the expansion of capitalistic modes of production (Brenner, 2001: 594; Marston, 2000: 227). Brenner claims that the "geoeconomic project of neoliberalism" which goes hand in hand with processes of commodification has transformed the scalar organization and scales of "socio-political regulation" (Brenner, 2001: 594). State interventions facilitated privatization and scalar restructuring, which have territorial and tangible outcomes inscribed in landscapes (Marston, 2000: 221–227). Based on Brenner, I use the concept of scale in a process-based sense, focusing on the political processes (e.g. decentralization of forest governance or the construction of a new transnational scale of forest governance) which lead to the production of new scalar configurations and which alter existing scalar configurations and scalar hierarchies (Brenner, 2001: 600). In Indonesian forest landscapes, the so-called sector laws (e.g. Forest Law, see Chapter 3 for more detail) aimed to strengthen the control of the national government over natural resources and promote investment in natural resource exploitation by establishing a concession system. In particular, the Forest Law had far-reaching territorial consequences, challenging established modes of production (e.g. shifting cultivation practices conducted by customary communities) and scales of socio-political regulation (Elmhirst, 2001; Peluso et al., 2008; Pye, 2012; Rachman, 2011: 30–36; Towers, 2000). A new national concession system, for instance, can be considered as a new scale regulating access to land for the purpose of facilitating forest exploitation and agro-industrial estates. This new scalar structure restricts access for those actors that are not able to establish relationships with national authorities. Changes in the scalar structure might have tangible outcomes in landscapes, for example a transition from shifting cultivation to commercial large-scale oil palm plantation estates.

Second, the politics of scale literature seeks to explain dialectical relationships between structure and agency (Marston, 2000: 220; Towers, 2000: 26). Scales may also structure the livelihoods of actors (Wissen, 2008: 20). But actors may also actively change the configuration of existing scales and seek to produce new scales. Neil Smith (1992) stresses that human agency and social and cultural practices also contribute to scale production. He conceptualizes the politics of scale as the frictions and contestations within scales and between scales (Marston, 2000: 228; Smith, 1992: 64). Scale, as he understands it, "[. . .] both contains social activity, and at the same time provides an already partitioned geography within which social activity takes place" (Smith, 1992: 66). In

Smith's reading of scale, social conflicts take place on more than one scale. Especially subaltern groups, marginalized at the local scale, might seek to "jump" (Smith, 2008: 232) to a more promising scale. Scale jumping, Smith argues, is used as a resistance strategy that might facilitate alliances between actors or that might provide access to resources or to political decision making at higher or lower scales. Successful actors are consequently those who are able to choose the appropriate scale of political struggle for achieving their interests, this might include the deconstruction of scales serving the interests of political opponents (Smith, 2008: 232). Active scale choices can be used in order to include or exclude actors from access to political resources, natural resources and land (Lebel et al., 2005: 1).

Third, scale theorists focus on the social and ecological consequences of scale construction or reconstruction. As a geographical construction scales divide socio-material natures (Swyngedouw, 2010: 12). Consequently, scales can be understood as particular spaces containing social processes and biophysical processes (ibid). For instance, the construction of a REDD+ project as a new scale of forest governance may enhance forest protection but may also exclude certain actors, e.g. shifting cultivators and logging companies, from participating in negotiation processes concerning access and control of the project area. George Towers (2000: 26), drawing especially on Soja, Lefebre and Brenner, extends the discussion and argues that "the social production of space invests the landscape with meaning and regulation [. . .]" dividing social and biophysical processes within "[. . .] landscapes into scales of meaning and regulation". Both scales are produced through social struggle and may overlap spatially and institutionally (ibid.). Forests in Jambi, for example, can be considered as one specific occurrence of "second nature". They have been transformed through specific modes of production entangled with modes of social-spatial organization. In pre-colonial times, the most relevant modes of production were shifting cultivation and the gathering of non-timber forest products. These practices transformed nature and produced (cultural) landscapes invested with distinct scales of meaning, e.g. community forests as hunting grounds, community forests as spaces for shifting cultivation, and community forests as sources of non-timber forest products. Dutch colonization introduced new scales of meaning constructing Jambi's forest as a source of colonial wealth and linking Jambi's forests with the colonial administration in Jakarta and the Dutch government in the Netherlands. Different conceptualizations of nature and landscape lead to competing scales of meaning. The specific outcomes, e.g. Jambi's forest as a source of colonial or of community wealth, might contradict each other, inducing social conflict. Conservationists may refer to the particularities of a specific landscape, e.g. a habitat of the endangered Sumatran tiger or the carbon storage capacity of peat swamp, and consequently construct a scalar narrative creating a new scale of meaning (Hein et al., 2016: 381; Kelly, 1997; Swyngedouw, 2010; Towers, 2000). In contrast, indigenous groups construct scales of meaning based on their ancestral lands that offer alternative boundaries to legitimize their presence in a landscape (Hein et al., 2016: 381).

Territory, property and authority

Territory, property and authority are three deeply entangled social relations. There is a vast literature on the three concepts (Hall, 2013; Paasi, 2003; Peluso and Lund, 2011; Sassen, 2008). I will use them in the following ways. I consider *territory* as a socio-spatial relation linking space (in most cases land) and identity (Jessop et al., 2008; Hall, 2013). However, in contrast to scale, I consider territory rather as a flat non-hierarchical concept. *Property* can be defined as a legitimate social relation to objects of value (MacPherson, 1978). I define *authority* as a "[. . .] a specific form of power exercised publicly and legitimated with reference to the state" (Lund, 2008: 7) or to pre-existing authorities, such as lineage leaders or feudals controlling imagined or historical territories. In the following paragraphs I will explain the three concepts in detail.

Following Ansii Paasi (2003: 110), I consider "[. . .] territories as a social process in which social space and social action are inseparable". Territory as scale is socially constructed (ibid. 110). Paasi has identified a number of crucial elements that "make" a territory: a "process of naming", the creation of symbols such as flags, physical or symbolic border demarcation, institutions (e.g. state, parastatal or indigenous authorities), and day-to-day social practices that reproduce and internalize territoriality (ibid.). Consequently, territorialization is a process that goes beyond the production of space. It involves rule making, claims and sanctions (Peluso and Lund, 2011: 673). The concept also extends beyond property, thus beyond claiming land. It clearly includes an emotional connection to land which is important to identity formation, and a political authority which governs not only access to land use but also other aspects of social life (Hall, 2013: 11). Territories can be considered as "power relations written on land" (Peluso and Lund, 2011: 673). Territorialization goes hand in hand with the making of boundaries. Boundaries are lines of in- and exclusion (Paasi, 2003: 113). They separate resources of value from economically, ecologically or culturally less relevant places. In the context of protected area formation, this often implies the exclusion of people from land and other income sources (Hall et al., 2011; Hein and Faust, 2014).

Property as territory is a contested concept in social science, heavily loaded with different ideologies, ongoing and century-old philosophical discussions on its meaning and its function, intrinsically linked to the constitution of social identity and often perceived as a fundamental legal basis of societies (Lund, 2008: 3; von Benda-Beckmann et al., 2009: 1–2). The understanding of property in Western societies is dominated by the argument that formalized private property is a fundamental requirement for the efficient use of (natural) resources and for "proper" market exchange (van Meijl and von Benda-Beckmann, 1999: 6). According to Indonesian state ideology, property should contribute to development, market exchange and welfare (von Benda-Beckmann and von Benda-Beckmann, 1999: 30). In contrast, according to many indigenous ideologies such as *adat*, property should support and balance the livelihoods of the community and those within it (ibid.).

For John Locke property is a transformative product that emerges out of combining labor and nature (Macpherson, 1978: 18). Locke argues that a contract between the state and individuals guarantees property (van Meijl and von Benda-Beckmann, 1999: 2). Property refers to access to or physical possession of material objects of value and the ability to benefit from these objects based on enforceable rights (Macpherson, 1978: 3). Property can only be considered as such if a legitimate public authority sanctions it, and vice versa. Consequently, property is clearly linked to authority and territory but also to different scales of governance regulating, legitimizing and defining property rights. A central element of public authority is legitimacy. An authority, for example a village head allocating rights to natural resources, can only be considered as such if villagers show a minimum of voluntary compliance with rules imposed by the village head (Alagappa, 1995: 23; Lund, 2008: 7; Sikor and Lund, 2010: 1). According to Muthiah Alagappa (1995: 14, 31), constituting elements of legitimacy are shared norms and values, including the belief in sanctity and traditions, law (procedural legitimacy), the ability to control territory and power. Local public authorities, such as village heads in Indonesia, draw on the notion of the state to achieve legitimacy. They gain legitimacy through using symbols and languages of the state and through referring to the legal system of the state (Lund, 2006: 687).

What members of society consider to be legitimate changes over time and is subject to historically contingent and continuous struggles about concepts and truths within society (Sikor and Lund, 2010: 6). Struggles over the constitution and legitimacy of public authority and the legitimacy of a specific scale of regulation have direct influence on the legitimacy of the property relations in place. National laws, regulations and policies structure access and property relations. Local public authorities follow especially those policies and laws that support their own interest (Lund, 2008: 4). National laws and policies are not necessarily fully implemented locally or fully obeyed by local public authorities and by the local population. But they often structure the agency of local actors (ibid.). The relevance of a specific national legislation in a local setting depends on the power structure in the political arena (ibid. 135). The result is "[. . .] neither coherent policy implementation nor complete disregard of law and policy" (ibid.). Accordingly, what is perceived as legal or illegal is not only the result of changing laws, it is the outcome of what powerful local actors consider as appropriate (ibid. 19). When legitimizing specific local activities, e.g. the issuance of a village–scale land title[1] or of a forest conversion permit, local public authorities might translate or transform fragments of national regulations and policies (e.g. land laws or land-titling policies) which correspond to their objectives into locally relevant rules (Kunz et al., 2016).

The legitimacy of public authorities in local politics is in many cases characterized by "[. . .] endless chains of reference to bigger authorities" (Lund, 2006: 693), and consequently has a direct scale component. Different public authorities entangled with different scales might compete over legitimizing property rights for the very same piece of land. To strengthen legitimacy, public authorities seek to gain additional recognition from other institutions (Lund, 2008: 2), in many cases from institutions at higher scales of governance. Political scales are relevant as reference points for local public authorities. However, local public authorities actively produce and reproduce scales (e.g. village scale), as well as employ scalar strategies, such as scale jumping, to achieve their interests. Public authorities such as village heads, district heads, provincial governments and the national government stabilize scalar configurations through regular interaction with their citizens (and vice versa), for example by enforcing land claims (Lebel et al., 2005). They channel social interaction and stabilize the social production of space and scales (Towers, 2000: 26). Access to specific public authorities on higher or lower scales is an important factor in explaining socially differentiated abilities to benefit from resources (Leach et al., 1999: 233).

In particular, societies with plural land tenure systems have nested and plural legal authority arrangements (legal pluralism) with unequal ranges of validity and unequal abilities to enforce claims. Claims backed by high-level administrative authorities may have greater legitimacy than claims backed by a village official or vice versa (Sikor and Lund, 2010: 6f). In frontier regions, with their confusing and dynamic institutional landscapes, social identity is a key factor shaping the ability to access public authority and to benefit from resources (Hein and Faust, 2014: 23; Rhee, 2009: 53; Ribot and Peluso, 2003: 170). Ethnicity and kinship shape patron–client linkages and permit privileged access to state officials and, consequently, to formal or semi-formal processes facilitating resource access (Rhee, 2009; Ribot and Peluso, 2003). Ethnicity is context dependent (Wimmer, 2008: 977) and determines affiliation to groups with specific customary arrangements permitting resource access for their members. Classifications such as *putra daerah* (child of the region) or "first comers" and "late comers" serve as ethnic markers and as factors influencing access to natural resources and political power (Lund, 2008: 16; Rhee, 2009: 43).

A few notes on the transnationalization of the state and on market-led environmental governance

Recently, a number of scholars have started to strengthen political ecology by theorizing the state and its role in socio-ecological conflicts (e.g. Brad, 2016; Hein et al., 2018a; Ioris, 2014; Pichler, 2014, 2015). I argue that the state or, in broader terms, different "notions" of the state have an important role in agrarian conflicts, in the process of legitimizing property rights to land and in developing and enforcing environmental law. However, the state is not a homogeneous actor. Ulrich Brand and Christoph Görg (2003: 226), following the Marxist scholar Nicos Poulantzas (1978), conceptualize the state as a "[. . .] power-based social relation which creates, in the form of apparatuses, a materiality which by itself is full of conflicts and contradictions". For Poulantzas, the state consists of the different state apparatuses that "[. . .] organize the specific relations between the ruling classes" (Demirović, 2011: 43). State agencies on different scales, ministries and state administrations such as the forest service and the national land agency, represent the different functions of the state and the division of labor within the state (Poulantzas, 1978: 155). However, different policies developed and implemented by these different apparatuses (such as land tenure regulations and environmental laws) are neither unitary nor coherent (Demirović, 2011: 44; Hein et al., 2018a: 4–5). Different apparatuses mediate between competing actors and reflect their competing interests. In consequence, they often develop contradictory state strategies. For example, the agricultural agency may push agricultural expansion while the forest agency is expanding protected areas. Ambiguous state strategies often induce environmental conflict and reflect contradictions and compromises among powerful groups in society (Demirović, 2011: 44; Hein et al., 2018a: 4–5). The state reflects "societal relationships" and is part of society. In this sense, "[. . .] the 'state' and '(civil)' society [are] formally separate, at the same time they form a contradictory unity" (Brand et al., 2008: 35).

Broader restructuring of global capitalism (Brand et al., 2011; Demirovic´, 2011: 51, Robinson, 2001: 164; Wissen, 2011: 234) and the reorganization of political power in Indonesia have transformed the spatial organization of the state and societal relationships with nature (Aspinall, 2013: 37–39; Hadiz, 2001: 144–145; Hein et al., 2018a). For this book, two aspects of state transformation are relevant: first, the transnationalization of environmental governance and the formation of transnational state apparatuses; second, increased competition between states, in other words the formation of international and transnational competition states (Behrens and Janusch, 2012; Hirsch and Kannankulam, 2011: 26).

First, the "theory of a transnational state" considers the state as a heterogeneous multi-layered and multi-centered network of actors that consists of national actors, supra-national institutions such as UN agencies, the World Bank, IMF and ASEAN, and transnational lobby groups, such as environmental NGOs, private forest carbon standards and business associations (Demirović, 2011; Hein et al., 2018a; Robinson, 2001). Robinson (2001: 158) argues that "[. . .] economic globalization has its counterpart in transnational class formation and in the emergence of a transnational state that has been brought into existence to function as the collective authority for a global ruling class". In the realm of environmental governance, the mentioned actors that constitute the apparatuses of the transnational state have established a set of transnational and international rules that provide a framework for global market-oriented solutions to maintain economic growth in the context of multiple environmental crises (McAfee, 2012b: 26). Global market environmentalism, including mechanisms such as REDD+, require transnational and national state power (Pellizzoni, 2011: 796), new transnationalized authorities and the transnationalization of environmental regulation (Heyvaert, 2017) to operate. Especially climate governance has experienced a significant shift towards transnational network forms of governance where private actors (such as carbon brokers and carbon standards) are becoming increasingly relevant (Bulkeley and Newell, 2015; Chan et al., 2015, 2016). At the same time, the emergence of global peasant movements like La Via Campesina and the Asian Peasant Coalition have transnationalized resistance against neoliberal agrarian policies such as market-led land reforms (Borras, 2008) and market-based conservation instruments (Cabello and Gilbertson, 2012: 175; Hein and Faust, 2014: 23).

Second, in Indonesia and in other emerging economies, the interventionist development state has been transformed into the "national competition state" (Hirsch, 1995; Hirsch and Kannankulam, 2011) supporting the privatization, commercialization and transnationalization of natural resources (Pye, 2012: 202). In Indonesia, this transformation did not occur in a unidirectional fashion, reflecting tensions between protectionists and world-market-oriented actors (Ufen, 2002: 120). Andreas Ufen (2002: 124) has identified three phases of state transformation since Suharto came to power: first, a period of economic liberalization between 1965 and 1974; second, a period of petro-dollar-financed interventionist development policies between 1974 and 1983; and, third, again and still ongoing, a period of economic liberalization associated with deregulation, privatization, decentralization and market opening (ibid.).

National competition states are increasingly considering the opportunities of carbon, biodiversity and conservation markets and seek to provide perfect framework conditions for private investors, and for the interventions of transnational NGOs and international donors (Brand and Görg, 2003; While et al., 2010). Biodiversity offsets, genetic resources, privatized conservation and forest carbon offsets constitute the "ecological phase" of capitalism (Escobar, 1996: 326). They are becoming increasingly relevant to maintaining economic growth but also accessing climate finance (e.g. Green Climate Fund) and carbon trading instruments. In Indonesia, the first market-based conservation policies delegating protected area management to private actors emerged in the 2000s, permitting private conservation concessions and payment for ecosystem service projects. The first regulations permitting private actors to run REDD+ offsets were issued in 2008 (Hein, 2013b; Hein and Faust, 2014; Walsh et al., 2012a).

Based on the above-mentioned spatio-institutional transformations and on the work of Maureen G. Reed and Shannon Bruyneel (2010: 651), I derive three rescaling processes relevant in the context of REDD+ implementation in Indonesia:

- the up-scaling of state functions towards international and transnational state apparatuses (e.g. UNFCCC, FCPF);
- the down-scaling of state functions towards regional state apparatuses and local communities (e.g. local governments, community-based conservation projects);
- the scaling-out or delegating of state functions towards non-state actors (e.g. conservation companies running REDD+ projects, transnational carbon standards certifying forest carbon offsets).

Conceptualizing power and resistance

Power is considered as an important explanatory factor for differences in the abilities of actors to access land and property and to develop multi-scalar resistance strategies (Corbera and Brown, 2010; Koch et al., 2008; Nuijten, 2005; Ribot and Peluso, 2003; Rodriguez de Francisco and Boelens, 2014; Wynberg and Hauck, 2014). Different schools of thought have influenced political ecology and the literature on the politics of scale. Post-war French Marxism, Gramsci's understanding of hegemony, post-structuralism based on Foucault and realism have been picked up and combined by different political ecologists and scale theorists (e.g. Bryant, 2001; Ekers et al., 2009; Forsyth, 2008; Mann, 2009). In this section, I will first take the notions of a number of great theorists as a point of reference for a brief review of the different power concepts used in the politics of scale literature and in political ecology. Thereafter, I will argue for a three-dimensional conceptualization of power based on the work of Steven Lukes (2005) and John Gaventa (Gaventa, 1982)

- For Michel Foucault (2006: 14–15), power is an ensemble of mechanisms and procedures, which is inherent in all social relationships. In his understanding, power is not attached to people, institutions, or class (Balan, 2010: 38; Ribot and Peluso, 2003: 156). Furthermore, Foucault links power intrinsically to knowledge and discourse (Gaventa, 2003).
- Antonio Gramsci uses the term *hegemony* for describing unjust power relations between different social actors. Hegemony, as the dominance of one group over another, is achieved through social relations of coercion and consent (Karriem, 2009: 317) and especially through "[...] active and moral and intellectual leadership" (Ekers and Loftus, 2008: 702).
- In post-war French Structural Marxism, power is defined as the ability of social groups (e.g. classes) to achieve their (class-specific) interests. Power, according to Nico Poulantzas (1978), is relational and not quantifiable. He argues that power emerges from a relational system of material positions that different social actors can hold (ibid. 136).
- In realism, power is conceptualized as the ability to achieve one's objectives (Keohane and Nye Jr, 1998: 86). The ability to achieve objectives is determined by the ability to control the necessary resources (e.g. financial resources, weapons, organizational strength) (ibid.).

Early scale theorists such as Smith (2008, 1992) argue explicitly that power asymmetries between specific actors, e.g. between classes, are inherent in capital–labor relations and are reflected in the socio-spatial organization, or – in other words – in the scalar structure. Many authors focus on the consequences of power asymmetries as explanations for scalar configurations. Most of the authors share the following arguments: scales are spatial manifestations of power, changing power relations may influence the scalar structure, and scale-power relations are dialectical (Meadowcroft, 2002; Swyngedouw, 2004; Zulu, 2009). Some authors frame power as a capacity and as based on material resources, e.g. as the ability of certain actors to accomplish certain activities (Allen, 2003: 97; Ekers and Loftus, 2008: 701; Swyngedouw, 2004: 17; Zulu, 2009). Lebel (2005) argues that powerful actors have the ability to influence social and political processes on different scales. Authors such as Swyngedouw and Zulu use rather critical realist framings of power and implicitly combine these with an understanding of power akin to Gramsci's concept of hegemony.[2]

Political ecologists argue that unequal power relations are an important explanatory factor for uneven access to natural resources, including access to land. (Blaikie, 2012; Bohle and Fünfgeld, 2007; Bryant, 1998; Forsyth, 2008; Rodriguez de Francisco and Boelens, 2014). Nevertheless, in many cases power is not explicitly conceptualized and understandings of power within political ecology have changed significantly over time. Early work, often referred to as *first phase political ecology scholarship* (Bryant, 1998; Forsyth, 2008), links structural Marxism with thoughts on the emergent environmental crisis (Forsyth, 2008: 758) and consequently draws on rather Marxist definitions of power (e.g. the ability of classes to achieve their class-specific interests).

The second and third phase of political ecology has various theoretical references. For some scholars, structural Marxism remained important while others increasingly shifted towards Foucault, Gramsci and others (e.g. Bryant, 1998; Ekers et al., 2009; Peet et al., 2011). A political ecology based on Gramsci inquires into, for instance, "[. . .] how hegemony is achieved through particular spaces and natures" (Ekers et al., 2009: 288). In the "Theory of Access", Jesse Ribot and Nancy Peluso define power in line with Foucault as "[. . .] embodied in and exercised through various mechanisms, processes and social relations – that affect people's ability to benefit from resources" (Ribot and Peluso, 2003: 154). In addition, they have a rather critical realist understanding and argue that "ability is akin to power". They cite Steven Lukes (1986: 3 cited in ibid. 155) arguing that power is defined "[. . .] as the capacity of some actors to affect the practices and ideas of others" (ibid.).

Three-dimensional power

This short literature review shows that different scale theorists and political ecologists have different understandings of power. Likewise, many scholars refer to power but do not explicitly conceptualize or analyze it empirically. However, I argue that political ecological research would benefit from more explicit engagement with power theories and from empirical analysis focusing on power asymmetries and their root causes. I believe that Steven Lukes' (2005) and John Gaventa's (1982, 2006) "three dimensional" view on power, as suggested by Jean Carlo Rodriguez de Francisco and Rutgerd Boelens (2014), provides a number of very useful "tools" to analyze power in its different forms.

Lukes (2005) and Gaventa (2006) distinguish three different types of power which have to be understood in relation to different socially produced political arenas and scales (Gaventa, 2006: 25; Rodriguez de Francisco and Boelens, 2014: 353). The three dimensions of power reflect complementary ways of thinking about power. They are also used at least implicitly in the politics of scale and explicitly in some of the political ecology literature (Ribot and Peluso, 2003; Rodriguez de Francisco and Boelens, 2014; Rodríguez de Francisco et al., 2013).

According to Gaventa (1982, 2006), *visible power* (the first dimension of power) refers to material resources, capacities, organizational strength and to participation in decision-making processes, and is akin to realist understandings of power. Visible power "[. . .] may be understood primarily by looking at who prevails in bargaining over the resolution of key issues" (Gaventa, 1982: 14). In other words, visible power[3] refers to the capacity of actor A to get actor B to do things that are against his own interest (Lukes, 2005: 16–17). Visible power includes economic resources, such as financial assets or land and political resources, e.g. the number of members of a political organization.

Hidden power (the second dimension of power) refers to the "rules of the game", e.g. according to Barach and Baratz (1970, cited in Gaventa, 1982: 14) to "[. . .] a set of predominant values, beliefs, rituals and institutional procedures that operate systematically and consistently to the benefits of certain persons and groups at the expense of others". Hidden power also refers to the ability of actors to set the community agenda and to exclude certain actors, and to mechanisms ensuring compliance with rules (Gaventa, 2006: 29; Lukes, 2005: 21). As Lukes (2005: 111) expresses it, hidden power refers to the " [. . .] power to decide what is decided".

Invisible power (the third dimension of power) is akin to Gramsci's concept of hegemony. However, hegemony for Lukes (2005) refers to unconscious and internalized domination and subordination and to active and coordinated strategies to achieve or to resist domination. Invisible power points to forms of power that "[. . .] prevent people, to whatever degree, from having grievances by shaping their perceptions, cognitions and preferences in such a way that they accept their role in the existing order of things, either because they can see or imagine no alternative to it, or because they see it as natural and unchangeable, or because they value it as divinely ordained and beneficial" (Lukes, 2005: 28). Invisible power importantly shapes how marginalized social actors perceive and accept the dominant social production of nature and space (Gaventa, 1982: 16–19).

Gaventa (2006: 25) argues that Lukes' three power dimensions are interrelated sets of dynamic relationships rather than static categories which operate across scales and spaces. Gaventa acknowledges that power and space are intrinsically linked. Powerful actors or successful resistance movements have to apply different forms of power across spaces and scales to maintain their interests or to change existing power relations significantly.

In order to identify power empirically (Table 2.1), and especially to identify the role power has for the ability of different actors to access land and property, this study builds on and extends a methodology developed by Gaventa (1982: 20–32). Visible power can be identified by investigating: the organizational strength of a specific organization (e.g. members), the material resources (e.g. land, financial resources), who prevails in formal decision-making and who holds formal or customary functions (e.g. village head, hamlet head, customary leader) (ibid.). Hidden power can be identified by investigating: non-involvement in decision-making (e.g. by investigating alternative developments facilitated by power-shifts), resistance activities (e.g. land occupations), communication and the socialization of subordinate actors (e.g. language used by

Table 2.1 Overview of different power dimensions (based on Gaventa (1982, 2006) and Lukes (2005))

Type of power	Explanation	Criteria for identifying different types of power
Visible Power (first dimension)	(Political) resources, organizational strength, who participates and prevails in decision-making, the power to affect others directly.	Asset base, e.g. land; financial resources, social capital, e.g. membership in groups; participation in formal decision-making bodies, e.g. village meetings; the holding of a formal or customary function, e.g. *Kepala Desa, Kepala Dusun, Ketua Adat.*
Hidden Power (second dimension)	Rules of the game, values, agenda setting, exclusion of specific actors by rule making, mechanism of compliance, coercion and exclusion.	Involvement and non-involvement in decision making, communication and socialization of subordinate actors, identification of mechanisms of exclusion (non-decision making), ethnicity and social identity, existing state and customary regulations and the extent to which they privilege specific actors over others, altered rules of the game.
Invisible Power (third dimension)	Construction of meanings, myths, language, moral and intellectual leadership.	Based on interpretation, e.g. what would be of interest for a certain actor, scales of meanings constructed for legitimizing benefits for elites, internalization of subordination, acceptance of authority relations, internalization of development and backwardness narratives of the New Order era, privatization and commodification of land and conservation.

subordinate actors), and the historical development of specific legal orders (e.g. established by force in the context of colonization) (ibid). The identification of invisible power is mainly based on interpretation, as Gaventa argues (1982: 29), considering, for instance, what would have been of interest to an actor in a specific situation. Furthermore, the identification of beliefs of inferiority by actors and the explicit acceptance (e.g. through statements) of a certain legal order that acts against their interest might further support the identification of invisible power (ibid. 31).

Resistance and three-dimensional power

Caouette and Turner (2009b: 9) argue that "[. . .] conceptualizations of resist-ance are situated within understandings of power; power being comprised of the relational interplay of dominance and subordination". Gaventa (1982: 23) has pointed out that resistance or, as he puts it, "[. . .] rebellion may develop if there is a shift in the power relationships – either owing to loss in the power of A or gain in the power of B". Peasant resistance occurs in open and rather collective forms or in hidden and rather individual forms (Chin and Mittelman, 1997; Turner and Caouette, 2009). Both forms are spatial and territorial practices (especially in the case of land conflicts), challenging the power constellations in place but also the pre-existing scalar structure and pre-existing scales of meaning and regulation (Moore, 1998; Turner and Caouette, 2009; Towers, 2000). Hidden and open forms of resistance have to be understood in relation to the conceptu-alization of power introduced above and should be treated as different categories in social science, to be specific as dynamic, interrelated and overlapping.

Hidden resistance, in James C. Scott words "every day peasant resistance" or "infrapolitics", comprises relatively safe and silent practices (Chin and Mittel-man, 1997: 31; Scott, 1989: 34). "Every day peasant resistance" is characterized by resisting without openly contesting the existing political order. It is rather negotiated in the informal sphere, e.g. household or community, and does not openly challenge hegemony (Chin and Mittelman, 1997: 31). Peasants prefer relatively safe resistance strategies to avoid open conflict with actors that are more powerful (Scott, 1989: 34–35). Scott (1989: 35) and Turner and Caouette (2009: 11) argue that when power constellations change, individual and hidden peasant resistance can turn into larger resistance activities, e.g. silent encroach-ment on plantations or protected areas can develop into larger organized and open land occupations.

Power constellations change, argues Gaventa (1982), when subordinates, or in his words "the powerless", are able to challenge all three dimensions of power (ibid. 24). To develop material resources (visible power), the powerless have to defeat invisible and hidden power (ibid.). To overcome invisible power, the powerless have to develop an understanding of their subordinate role in society. Moreover, they need to develop strategies for changing the existing political circumstances and intrinsically linked power constellations (ibid. 21). Strate-gies and the development of political objectives and issues of concern allow political mobilization, thus overcoming hidden power. In many cases, powerless actors employ clandestine strategies, such as ignorance, smuggling, sabotage, or silent encroachment on state forest land (Gaventa et al., 2011; Scott, 1989: 34). The various strategies might then allow the development of visible power and engagement in open resistance (Gaventa, 1982: 21–24). Peasant groups such as Serikat Petani Indonesia (SPI) have developed the means to employ open resistance strategies, e.g. larger-scale occupation of state forest land and demonstrations (see Chapters 3 and 5 for more detail).

Resistance as a spatial strategy might also involve "scale jumping". The term *scale jumping* has been picked up by many authors for describing attempts to shift the political struggle of actors that are marginalized at a specific scale to higher or lower scales[4] (Hein and Faust, 2014: 24; Smith, 2008: 232; Zulu, 2009: 687). Gaventa (2006: 31) argues that contemporary resistance movements need to build up strategic alliances with actors operating on different political scales to be successful. Scale jumping, in other words the spatial expansion of protest to higher scales, has been facilitated by the transnationalization of governance and growing interrelations between formerly separate national policy fields (Caouette and Turner, 2009a; Smith, 2008; Swyngedouw, 2000).

Key arguments

Beyond understanding the root causes of the agrarian conflicts challenging REDD+ implementation, this book aims to advance our conceptual and theoretical understanding of emerging agrarian conflicts in the context of transnational conservation interventions. It aims to contribute to a "political ecology of transnational agrarian conflicts in the context of REDD+".

The different elements of the framework help to investigate differing but interrelated empirical phenomena. Political scales, for example, demarcate arenas of political contest (Smith, 1992: 66). In line with Swyngedouw (2010), I assume that questions of access to natural resources (including land) can be explained by analyzing the socio-spatial configurations of scales. I argue that rescaling processes in the course of state transformations (e.g. colonization, nationalization, decentralization) structure the abilities of different social actors to access property rights. Thus, I consider scales as arenas of political struggle which are not fixed and which are linked through actor networks and scale-jumping strategies (Bulkeley, 2005; Flitner and Görg, 2008; Swyngedouw, 2010). Social conflicts could change the scalar configuration (e.g. deconstruct pre-existing scales, widen existing scales or construct additional scales) and the material and biophysical content of a specific scale. Powerful groups are able to actively choose or alter the scale of regulation to achieve specific interests. Subaltern groups seek to jump to higher or lower scales or to establish networks across scales to achieve their interests (Perreault, 2003: 65; Zulu, 2009: 695). Thus, scale and rescaling reflect the dialectical relationship between structure and agency. They structure and contain space and social practices within space and they are the outcome of social practices (Smith, 1992: 60).

The work of Towers (2000) provides the opportunity to analyze the social construction of scales of meaning and scales of regulation separately. This is of specific importance in the context of REDD+ and conservation debates where different scales of meaning such as carbon forests or customary forest overlap with local, national and transnational scales of regulation (Flitner and Görg, 2008; Lebel et al., 2005). Brenner stresses the central role of the state in producing hierarchical scalar structures (Brenner, 1997, 2001). Political scales in this sense also constitute and are the result of the division of labor among different apparatuses of the state. I argue that public authorities (such as the village head, the forest service and district government) and private actors (such as forest carbon standards) seek to control their respective scales of regulation. By legitimizing the property rights of peasants and companies, or by issuing a carbon credit, they regularly engage with their citizens and customers and construct and maintain a specific scale of regulation in the first place.

Public authorities have the ability to exercise power publicly. They often build on symbols and the legal texts of the state (Lund, 2008: 7). In order to establish legitimacy, in many cases public authorities refer to authorities at higher ends of the scalar hierarchy (Lund, 2006: 693) or seek to scale back (Reed and Bruyneel, 2010: 651) by making reference to past authorities and interrelated political scales (e.g. imagined lineage chiefs). The legitimacy of a public authority and of a specific scale of regulation is a matter of degree (e.g. continuum between strong and weak) and changes over time while new public authorities emerge. As in other countries, in Indonesia different public authorities, in other words different apparatuses of the state, might act in contradictory ways and thus might not reflect the will of apparently dominant groups in society. Scales of regulation constructed by public authorities and stabilized by interactions with citizens, e.g. through issuing land titles, might overlap or compete.

Especially in the context of rapid scalar restructuring (e.g. decentralization) scales of regulation may overlap leading to competing access and property relations. The transformation of the Indonesian state (the transformation from the colonial state, to the development state of the Suharto era, and to the competition state of the *Reformasi* era) has also changed the socio-spatial configuration of Indonesia's forest and land tenure governance significantly. In this context, Indonesia's forest apparatus has promoted foreign direct investment and the commodification of natural resources, land and very recently of ecosystem services (Nevins and Peluso, 2008).

Conservation territories, such as the Berbak Carbon Initiative or the Harapan Rainforest, and interrelated socio-ecological scales contain social and physical processes. Social conflict and changing power relations have an impact on the structure, content and spatial extent of conservation territories (Swyngedouw, 2010). The successful protest of an environmental movement might lead to an expansion of the borders of the Berbak Carbon Initiative changing the characteristics and meanings of the ecosystem and of the social relations contained by the project area. The expansion of the project may restrict specific land use practices, thus changing land use practices and social positions within the eco-system of actors. However, the very same ecosystem – or in other words, the very same space – might be experienced differently depending on the social position of an actor (Escobar, 1999: 5). As outlined above, for some actors a forest is an empty space "ready" for agricultural expansion whereas a REDD+ project developer might consider the same space as a site for the production of forest carbon credits.

Lastly but equally importantly, I argue that the ability to change scales of meaning and regulation and consequently the ability to access land and prop-erty is linked to the question of power. Social actors that are willing to change access and property relations have to make use of different forms of power across spaces and scales (Gaventa, 2006). Social actors have to rely on resources (visible power), have to have the capability to change formal and informal regu-lations (change the rules of the game, hidden power) and have to influence the "organizing ideology" (invisible power) (Alagappa, 1995).

Multi-sited qualitative research

After outlining the conceptual framework, one question remains: How to inves-tigate the political ecologies of REDD+ empirically? REDD+, and indeed most of the other relevant themes of this book such as property rights, forest governance and agrarian resistance, are negotiated and regulated on various political scales, e.g. at the village scale by the village head, by the district head or (in the case of forest land) by the Ministry of Forestry in Jakarta, or (in the case of REDD+) at UNFCCC conferences and in the headquarters of donor agencies. Consequently, the empirical research for this book was "multi-sited" and inspired by "deterritorialized" (Merry, 2000: 130) and "multi-sited eth-nography" (Marcus, 1995: 80). Multi-sited ethnography was first outlined by Marcus (1995) as a response to "empirical changes" (ibid. 80) in an increasingly globalized world. He argued, referring to Immanuel Wallenstein's world system theory, that research which is "embedded in a world system [. . .] cannot be [. . .] focused on a single site of intensive investigation" (ibid. 79f). In multi-sited fieldwork the so-called "global" is not external to the field of investigation, the global is rather part of the relationships constituting the field of research (ibid. 86). Multi-sited fieldwork is process-based and "[. . .] emerges from putting questions to an emergent object of study whose contours, sites and relationships are not known beforehand" (ibid. 86).

Multi-sited research, according to Marcus and Merry, should trace the net-works of interaction between actors across field sites. This provides the oppor-tunity to analyze peasants' resistance across scales and to investigate the attempts of actors to alter the scalar configuration or to identify scales that actors pref-erably use for political struggle. Multi-sited ethnography follows the "transla-tors and intermediaries" (Merry, 2000: 131). Translators and intermediaries are those actors that travel between field sites and link the different sites of social struggle and empirical investigation; this can include village heads, community representatives and members of NGOs (ibid.). Researchers conducting multi-sited research have different options to guide their investigations. Marcus (1995: 90–94) proposes following people, things (e.g. commodity chains), metaphors, plots, stories, lives or conflicts. This book seeks to follow conflicts.

Doing multi-sited fieldwork and selecting the main sites of investigation

The main sites of empirical investigation of this book are the Harapan Rainforest (*Hutan Harapan*/Forest of Hope) and the Berbak Carbon Initiative on the island of Sumatra, Indonesia (Figure 2.1). In addition, the book is based on qualitative research conducted along the lines of different networks of interaction linked to project implementation and resistance across field sites, namely the provincial capital of Jambi, the national capital Jakarta, the city of Bogor as an important hub for academia and conservation NGOs, UNFCCC climate change conferences, and the headquarters of donor agencies in Germany. Field trips to the project sites were conducted in 2012, 2013 and 2016. At the various field sites, the author has conducted (in Indonesia with the support of an Indonesian field assistant) semi-structured and open interviews, focus-group discussions and participatory observations. In addition to the different types of qualitative interviews as a primary empirical source, analysis focused on various types of documents such as Indonesian land and forest policies and laws, political strategies, NGO reports and scientific literature on REDD+ and land-use conflicts.

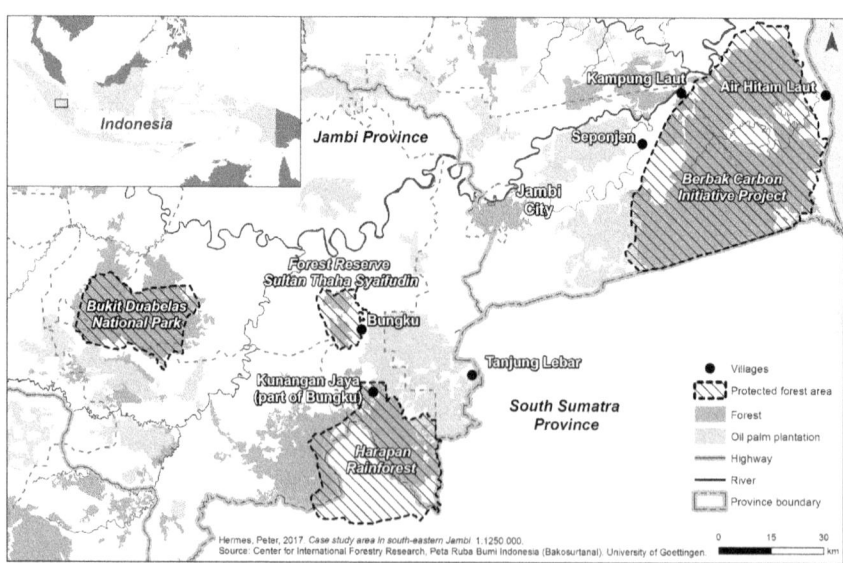

Figure 2.1 Map of Jambi and main sites of empirical investigation

Empirical research for this book was embedded in two research projects, namely the collaborative research center on "Ecological and Socio-economic Functions of Tropical Lowland Rainforest Transformation Systems" (CRC 990) in Jambi led by the University of Göttingen, and the research and advisory project on "Climate Change and Development" at the German Development Institute in Bonn. In addition to these institution-related reasons for doing research in Jambi, the province is of particular interest because Jambi became an Indonesian REDD+ pilot province in 2013, because of the engagement of the German International Climate Initiative (IKI), and because of land conflicts taking place within or at the margins of the REDD+ pilot projects and conservation initiatives.

When selecting the research villages within or adjacent to the two REDD+ projects, I built on the "follow the conflict" premise developed by Marcus (1995). Furthermore, sites with conflictive access and property relations, e.g. competing property rights systems, were preferred over other potential sites. With the support of researchers from the University of Jambi and the Agricultural University of Bogor and based on interviews with NGOs (mainly with environmental NGOs, environmental justice groups and peasant organizations), the staff of donor agencies and with village heads, I identified five research villages (Table 2.2) located within and adjacent to the REDD+ projects. Each village has its own specific conflict history, e.g. with oil palm plantation companies as well as a shared history of conflicts related to broader rescaling processes such as colonization and decentralization. Within the villages, I have focused in particular on conflicts that are affected or caused by the implementation of protected areas and REDD+ pilot projects.

Additional field sites are the transnational, national and local nodes of REDD+, land and forest governance. Nodes of the transnational REDD+ governance include, for instance, the ministries and governmental institutions involved in REDD+ and located in Bonn, Jakarta and Jambi City, UNFCCC conferences and the headquarters of environmental organizations (head office of Burung Indonesia in Bogor), peasant movements (e.g. SPI head office in Jakarta) and donor agencies (e.g. offices in Jakarta and Germany) (von Benda-Beckmann,

Table 2.2 Research villages, REDD+ and conservation projects

REDD+ conservation project	Research village	District
Harapan Rainforest	Bungku (including the hamlets of Kunangan Jaya I and II)	Batang Hari
	Tanjung Lebar (including the hamlets of Tanjung Mandiri and Pangkalan Ranjau)	Muaro Jambi
Berbak Carbon Initiative	Air Hitam Laut	Tanjung Jabung Timur
	Seponjen	Muaro Jambi
	Kampung Laut	Muaro Jambi

von Benda-Beckmann and Griffiths, 2005; Bulkeley, 2005; Flitner and Görg, 2008; Merry, 2000). These different locations are linked through decision-making processes and resistance strategies. They are part of competing and contradictive scales of meaning and regulation. Following von Benda-Beckmann and colleagues (2005: 9), I tried to follow "[. . .] the chains of interaction connecting transnational, national and local actors in multi-sited arenas of negotiation along with power relations that structure these interactions and are reproduced or changed by them".

The two REDD+ project sites and the various case-study villages

The area of the **Harapan Rainforest** project has 98,555 ha and is officially located in the district of Batang Hari and Sarolangun and in the neighboring province of South Sumatra (Figure 2.1). The north-eastern parts of the project area are claimed by the district of Muaro Jambi, but according to the maps of the Ministry of Forestry they are part of the Batang Hari district. The conservation company PT REKI runs the Harapan Rainforest project. The company is owned by a transnational NGO consortium consisting of

Figure 2.2 Picture of road sign of Bungku village and border signs of the Sultan Thaha Syaifudin Forest reserve

(Source: taken by the author, 2013)

Birdlife International, the Royal Society for the Protection of Birds (RSPB) and Burung Indonesia and has received funding from, among others, the German International Climate Initiative (IKI), the Danish Ministry of Foreign Affairs (DANIDA), the European Commission and Singapore Airlines (further information in Chapter 4). The project is implemented in an ecosystem restoration concession. As case-studies, I selected the villages of Bungku and Tanjung Lebar adjacent to the Harapan Rainforest according to the "follow the conflict" premise (Table 2.2).

The village of Bungku is located in the district of Batang Hari and in the sub-district of Bajubang north of the Harapan Rainforest project and south of the Sultan Thaha Syaifudin Forest Reserve. The village arose from a resettlement scheme for local Batin Sembilan groups in 1972 (Faust et al., 2013: 9; Hein, 2013b: 15). However, a number of pre-existent settlements in the area prove that the area has been populated since pre-colonial times (Zainuddin, 2013: 6). The village has approximately 10,215 inhabitants and a village territory of 77,000 ha (Faust et al., 2013). Land use in Bungku is dominated by oil palm and rubber cultivation. Shifting cultivation is still practiced by some households but is becoming less relevant. In any case, most of Bungku's village territory is either designated as state forest (e.g. part of the Harapan Rainforest project or of the Sultan Thaha Syaifudin Forest reserve) or is part of large plantation concessions.

The village of Tanjung Lebar is located in the district of Muaro Jambi and in the sub-district of Bahar Selatan. The center of the village is located north-east of the Harapan Rainforest concession. The village officially has 2876 inhabitants; the population is probably much larger, since official village data only exists for the hamlets located outside of the Harapan Rainforest (Polsek Sungai Bahar, 2011). The autochthon population of the village consists of Batin Sembilan, Melayu Jambi and Melayu Palembang. Tanjung Lebar dates back to the pre-colonial era. The official village territory has an area of 6500 ha, but again this figure does not include the area within the Harapan Rainforest claimed by the village. The village territory is fragmented, consisting of different dispersed and unconnected hamlets located between transmigration settlements and corporate oil palm plantations.

The **Berbak Carbon Initiative** is located in coastal Jambi, south of the Batanghari River delta and northeast of the provincial capital (Figure 2.1). It has a project area of 250,000 ha. The landscape is dominated by tidal peat swamps that are partly degraded because of settlement formation, the construction of drainage channels, forest conversion and almost annual peat fires (Claridge, 1994; Giesen, 2004; Hein et al., 2018a). The Berbak Carbon Initiative is listed as a REDD+ demonstration project by the Indonesian REDD+ agency (further information in Chapter 4). The project is a collaborative initiative by the Zoological Society of London (ZSL), the Jambi-based NGO Gita Buana, the Berbak National Park Agency and the Provincial Forest Service (Dinas Kehutanan Provinsi Jambi). Here I focused my investigations on the villages of Air Hitam Laut, Kampung Laut[5] and Seponjen (Table 2.2).

The village of Air Hitam Laut is part of the sub-district of Sadu and located in the district of Tanjung Jabung Timur. The village is located directly on the coast of the South China Sea, at the mouth of the river Air Hitam. The village territory has a size of 4320 ha (Alamsyah, 2004: 27). In 2012 the village had a population of 2328 inhabitants, according to the village secretary. Seventy-five percent of the population is Bugis and originally from South Sulawesi (ibid. 9). Land use is dominated by coconut plantations. Oil palm has been introduced recently. Air Hitam Laut was founded by Bugi seafarers from the Luwu district in South Sulawesi in 1965. Conflicts between the local population and the various forest authorities have been ongoing since the 1970s.[6]

The village of Kampung Laut is located in the district of Muaro Jambi. The village is located on the south-eastern riverbank of the tidally influenced lower course of the Batang Hari River. The village had 1573 inhabitants in 2012 and has a village territory of 12,000 ha. Melayu Jambi people are the dominant ethnic group. Oil palm, wet rice, rubber, cocoa and vegetables are the important crops. Fishing, swallow breeding and logging are other significant livelihood strategies. According to interviews with the village elders, the village was founded in pre-colonial times. In 2008, the Transmigration Agency of Muaro Jambi started to construct a transmigration settlement within the forest reserve, inducing land conflicts between various state apparatuses.[7] Kampung Laut is one of the pilot sites for a community reforestation scheme which is part of the community benefit package of the Barbak Carbon Initiative.

The village of Seponjen was founded in 1931 by the Melayu Jambi people (Kepala Desa Seponjen, 2013: 6). Seponjen is part of the sub-district of Kumpeh and belongs to the district of Muaro Jambi. Seponjen is located on both sides of the tidally influenced Sungai Kumpeh River, which is an arm of the Batang Hari River. Seponjen has a village territory of 16,000 ha and a population of 1315 people (ibid. 8). Wet rice and oil palm are grown in the flood-prone lower parts of Seponjen. On the higher lands rubber, cocoa, langsat[8] and durian[9] are cultivated. Since 1999, parts of the village territory have been designated as forest reserve. Seponjen is also one of the pilot sites for a community reforestation scheme which is part of the Berbak Carbon Initiative's community benefit package. The village is impacted by a number of land conflicts involving oil palm companies and the state forest agency.

Reflection, limitations and challenges

As Kim V. L. England stated, fieldwork is "[. . .] a dialogical process in which the research situation is structured by both the researcher and the person being researched" (England, 1994: 84). Multi-sited approaches also pose a number of challenges. Researchers using a multi-sited approach have to consider that positionality and power constellations (e.g. between researcher and participants) change across research sites. The researcher must be aware of the fact that his identity is constantly negotiated, affecting his ability to access relevant knowledge. Furthermore, the researcher must ensure that the questions he might ask have to be formulated in ways understandable for actors living in the different social fields under investigation. In the case of this piece of research, the different lingual contexts have posed an additional challenge. Interviews with German donor agencies and ministries were held in German, interviews at UNFCCC conferences and some of the expert interviews in Jakarta were held in English, and all other interviews (in Jambi city and in the research villages) were held in Bahasa Indonesia with the support of an Indonesian field assistant. Moreover, multi-sited research is probably not able to investigate the full complexity of local processes, since time at each location is much more limited than in single-sited research.

Research on conflicts, on illegal or semi-legal land occupations, and competing property right regimes creates a number of additional challenges as the researcher might be perceived as threatening the activities of the participants, as being part of the opponent conflict party or as a threat to the status quo (England, 1994: 85). The conservation company PT REKI and the Royal Society for the Protection of Birds (RSPB) as one of PT REKI's shareholders, for instance, argued from the very beginning that any research on land tenure would be impossible within the project area. A staff member of the RSPB argued that research in the conflictive areas of the Harapan Rainforest would exacerbate the conflict. Furthermore, he argued that conducting interviews would not be possible, adding that settlers' responses would give the impression of stronger tenure than was de jure the case (Manager of Royal Society for the Protection of Birds, 2012). The RSPB manager not only tried to hinder research, but he even directly questioned the validity of the statements that the settlers would make. PT REKI staff based within the project area argued similarly and only permitted research with communities that agreed to cooperate with the conservation company. Research in the conflict areas of Bungku and Tanjung Lebar was only possible because of permits and invitations from the relevant village governments, the settler groups and peasant organizations. Despite the permits from the village governments, the fact that PT REKI did not permit research in the conflict areas influenced my behavior at the field sites. Without an official permit from PT REKI, I felt unsafe, as someone acting "illegally", and tried to limit overnight stays in PT REKI's project area.

In the Berbak area the situation was different. The implementing agencies of the Berbak Carbon Initiative, such as the Provincial Forest Agency, the Berbak National Park Agency, ZSL and the sub-contracted NGO Gita Buana, supported my research project from the very beginning. The context of my fieldwork was consequently fundamentally different in the two areas, affecting my positionality in the field situations. In the Harapan landscape, my research was welcomed by village governments and informal settlers and rather rejected by the conservation company PT REKI. The informal settlers perceived my research as supportive for their concerns, especially those who were aware that the Harapan Rainforest project received funding from the German government. Settlers hoped that a German researcher would convey their concerns to the decision makers in Germany, and thus supported my fieldwork. In the

Berbak area, implementing agencies hoped that research from an "independent" third party might help them to solve ongoing conflicts.

In all of the villages, the village heads felt obliged to support my activities because of my formal letter of request from the University of Jambi. All the village governments supported my fieldwork, and most facilitated accommodation and access to interview partners. This formal support from the political elite of the villages facilitated my research but may have hindered my ability to cover marginalized actors within the political arena of my research villages. I encountered similar problems in the settlements controlled by the peasant movement SPI. SPI supported my trips to the settlement and all actors at the field site perceived me as an SPI guest. This probably discouraged local opponents of SPI from voicing criticism in interviews.

Additional challenges and limitations to my research are directly linked to my positionality and to power imbalances that structured interviews and fieldwork. As a researcher from Europe, I was highly visible in all field situations and attracted attention. As part of the "developed world" I was often asked about how trade-offs between development and conservation are solved in the "West", why Germany is supporting a conservation project in their region, or why forests in Jambi should be conserved to reduce Germany's greenhouse gas emissions. Questions of identity, of positionality and of being external to the community came up almost every day and were part of the field experience. The fact that a person from Europe was doing research in their village in Jambi made some villagers proud but also caused some individuals to be awestruck or afraid in interview situations and face-to-face communication. Furthermore, with my being a guest of the village government, some interview partners may have felt obliged to participate and to respond to questions despite having misgivings about discussing semi-legal land use or land access practices to outsiders. A general weakness of this study is that, as a male researcher, access to female household members and interview partners was difficult; therefore, only a few women participated in my interviews.

The next chapter is devoted to the politics of scale of Indonesia's forest and land tenure governance. It outlines the dynamic context of implementing REDD+ on the island of Sumatra and explains how customary authorities and village governments supported by peasant migrants successfully challenged the integrity of Indonesia's state forest.

Notes

1 Village heads in the various case-study villages issue documents used as land titles, but officially only the National Land Agency (Badan Pertanahan Nasional, BPN) has the authority to issue land titles (see Chapter 3 for more details on de jure and de facto land-titling procedures).
2 In their case studies, they show how different actors reshape the politics of scale of the Spanish waterscape and of community forest management in Malawi to consolidate legitimacy (Swyngedouw 2010, Zulu 2009).

3 Visible power is also akin to Max Weber's definition of power. Weber argues that men have power if they are able to achieve their objective against the will of others (Weber 1993: 20).
4 Keebet von Benda-Beckman (1981) developed the concept of "forum shopping" to explain the attempts of actors to choose between different conflict mediation institutions; she argues that their selection is based on the hope that the institution might support their interest.
5 Because of ethical considerations, I use a fictitious name of this village.
6 Interview with key informants in Air Hitam Laut, 28.09.2012, 30.09.2012, and with staff member of ZSL in Air Hitam Laut, 29.09.2012.
7 Interview with staff member of the sub-district administration (Kecamatan) of Kumpeh, in Sua Kandis, 02.09.2013.
8 *Lansium parasiticum*
9 *Durio zibethinus*

3 Rescaling of the governance of forests and land in Indonesia

The main assumption guiding this chapter is that broader rescaling processes in the course of state transformation – e.g. colonization, de-colonization, democratization and decentralization – changed the abilities of different actor groups to access land and property. Moreover, I argue that scaling back, in other words understanding historical scalar arrangements and associated land and forest tenure regulations, is a precondition for understanding land conflicts in REDD+ target areas (Eilenberg, 2015; Galudra et al. 2014; 2011; Howson and Kindon 2015; Mc Gregor 2010). The chapter shows that REDD+ and conservation initiatives are not implemented in a social and political vacuum. Land conflicts in the context of REDD+ implementation are embedded in historically contingent social and political dynamics.

In this chapter, I primarily investigate the dynamic context of REDD+ implementation in Indonesia, namely processes of state transformation, territorialization, intertwined reforms of forest and land tenure governance at different scales, agricultural expansion in the state forest territory, and a highly complex de facto land tenure regime. The different political regimes mentioned above facilitated the construction of specific scales of meaning and regulation. Some of them outlived the historical conditions that led to their construction. They are still inscribed in the landscape. Others have been deconstructed and replaced over time.

The chapter starts with a description of the pre-colonial and colonial forest and land tenure governance. It explains de jure land and forest tenure and subsequently scales down and investigates local histories of land governance, de facto land tenure and the attempts of sub-national governments, local authorities and indigenous groups to regain access to the state forest territory in the landscapes that later became part of the Harapan Rainforest and the Berbak Carbon Initiative in Jambi.

The history of Indonesia's forest and land tenure governance

Forest and land governance in Indonesia can be broadly divided into a pre-colonial phase, a colonial phase, a transitional phase of early independence, an autocratic and development-oriented phase (New Order) and a contemporary phase characterized by decentralization, democratization and (neo-) liberalization.

Each of the phases has its specific socio-spatial structure, was driven by different larger state projects and was influenced by broader narratives on how to use, manage and govern forest and land resources (Galudra and Sirait, 2009). The different phases had no abrupt end; some elements of each phase were persistent and influenced the subsequent phase. Indonesia's dualistic land governance system based on forest and agricultural laws, for example, has been inherited from the colonial period (Brockhaus et al., 2012: 32).

Indonesia as a nation-state was founded in the context of the colonial liberation war. No pre-colonial Indonesian nation state existed and consequently there was no pre-colonial archipelago-wide forest and land tenure governance. There is, however, no doubt that the various pre-colonial Sultanates and customary communities had rules for land and forest use in place. The next section discusses some of those known for Jambi and briefly introduces Jambi's pre-colonial history.

Jambi's pre-colonial history

In pre-colonial times, Jambi was one of a number of Sultanates on the island of Sumatra. Jambi's location close to the Strait of Malacca and the natural resources in its rich hinterland provided the basis for Jambi's ascent and its early involvement in international trade (Guillaud, 1994: 114). In the 18th century, Jambi was one of the most prosperous trade ports on Sumatra (Guillaud, 1994: 114; Locher-Scholten, 2004: 39). Forest products such as beeswax, resin, gum, rattan and timber were collected in the hinterlands and transported via the Batang Hari River and the Strait of Malacca to the outside world (Locher-Scholten, 2004: 37).

In 1852, Jambi had approximately 60,000 inhabitants (ibid. 36). In contrast to other Sumatran Sultanates, the population of the Sultanate of Jambi was relatively heterogeneous. The Malay population lived mainly on the banks of the Batang Hari River. The semi-nomadic Orang Laut lived along the coast (Locher-Scholten, 2004: 48). Jambi's hinterland was home to different nomadic and semi-nomadic ethnic groups. The Orang Rimba lived in the Bukit Duabelas Region. Batin[1] and Kubu[2] tribes settled on the upper courses of the main rivers and in the borderlands between Jambi and Palembang along the Batang Hari and Musi River tributaries (Andaya, 2008: 205; Locher-Scholten, 2004: 37; Steinebach, 2013a: 126).

Elisabeth Locher-Scholten (2004: 45) describes the pre-colonial Malay state of Jambi (the Sultanate) as a state without "[. . .] precise borders or a powerful central authority, their central government had no monopoly on the use of force, their rulers had no well-defined constitutional powers, and there was not a trace of popular sovereignty". The pre-colonial Sultanate had nothing in common with modern Western states, the "[. . .] ruler was more fluid in terms

of territory and structure, and sacral and symbolic in nature" (ibid. 45). Successful ruling was a result of balancing earth and divine and was indicated through
wealth and prosperity. In formal terms the Sultan was the highest authority of
the state but he ceded most of his tasks to the *pangeran ratu*,[3] to ministers and to
a council of nobles (ibid.).

Jambi's Sultanate had no uniform administrative structure such as charac-
terizes modern Western states. Jambi's administration reflected its heterogeneous population (Locher-Scholten, 2004: 48–49). In general, the Sultanate was
divided into different jurisdictions called *Kalbu*. Each *Kalbu* consisted of a number of settlements called *dusuns* (hamlets) (Hidayat, 2012: 29). The Sultan had
only direct authority over the Malay population, who had no tax obligations
but were responsible for law and order and for the security of the Sultan. The
Batin had to pay taxes and were responsible for border protection but were
relatively independent of the Sultanate. The Batin had their own leaders ruling
over their hamlets and over the surrounding forests (ibid.). Orang Laut and the
Kubu lived within the borders of the Sultanate but were not under the authority of the Sultan (ibid.).

Colonial forest and land tenure governance

Colonial forms of forest and land tenure governance in Indonesia date back to
the activities of the Dutch East Indian Company (VOC, *Vereenigde Oost-Indische
Compagnie*) in the 17th century (Galudra and Sirait, 2009: 525). The activities
of the VOC were mainly limited to timber extraction (ibid.). The first colonial
regulations were enacted for the island of Java in 1808 (Nurjaya, 2005: 38). This
first regulation defined any not privately owned forest areas as state domain,
established a colonial forest service accountable to the Colonial General Governor, and stated that the timber demand of the colonial government should
be prioritized. Furthermore, the regulation stipulated logging permits for local
communities (ibid.). Early colonial forest policies were aimed at facilitating
forest extraction (Galudra and Sirait, 2009: 525). During the mid-19th century,
colonial forest policies shifted towards conserving forests and maintaining associated environmental services such as water provision (ibid.).

Based on the belief that forests are the main determinant for the hydrological
balance of watersheds, colonial forest authorities argued that forest protection
was of colonial interest (Barr et al., 2006: 19; Galudra and Sirait, 2009: 525). The
main argument of foresters at the time was that forests act as sponges that are
able to assure water provision for irrigation agriculture during the dry season.
According to the dominant scientific discourse of the 19th century, deforestation – especially in the upper parts of watersheds – would significantly reduce
water availability in adjacent low lands (Galudra and Sirait, 2009: 525). The
argument that deforestation would disturb the hydrological balance was used
to legitimize the first resettlements, to legitimize the exclusion of local communities from forests and to legitimize the prohibition of shifting cultivation
(Galudra and Sirait, 2009: 530; Metzner, 1981: 47). Watershed protection was

constructed as a new colonial scale of meaning and regulation restricting access to land and forest resources to safeguard colonial interests.

Colonial forest policy was further developed during the mid-19th century. The first colonial forest law (*Boschordonatie*) for the islands of Java and Madura was issued in 1865. At the same time, the colonial government introduced a formalized concession system to facilitate forest exploitation (Nurjaya, 2005: 40). In 1870, the *domein verklaring*[4] declared that all vacant lands on Java and Maduro belong to the colonial state. In addition, the colonial forest agency initiated initial forest zoning activities (ibid.). In the same year, the colonial government enacted a second law, the *Agrarische Wet*. The *Agrarische Wet* aimed to regulate land ownership and promote private investment. It introduced a Western concept of private property, provided a framework for renting out land to private agricultural estate companies, and again declared all unused land to be the property of the colonial state (Gamin et al., 2014: 55; Nurjaya, 2005: 38; Szczepanski, 2002: 235). *Agrarische Wet* and *Boschordonatie* established a dualistic legal structure for land declared as forests and for land prescribed for agriculture which continued in post-colonial Indonesia and which is still reflected in contemporary Indonesian forest and agrarian laws (Indrarto et al., 2012: 36).

Colonial authority over land and forests on the outer islands[5] (e.g. Sumatra, Sulawesi) remained limited until the end of the 19th century. Colonial rule on the outer islands was mainly based on indirect rule and self-governance (Barr, 2006: 19). Dutch colonial authorities negotiated contracts with local Sultanates which forced them to accept the sovereignty of the colonial government and facilitated access to resources for Dutch and European companies (Locher-Scholten, 1994: 95; 1996: 140–141). Forests remained under the authority of the Sultanates, of allied elites and local communities. At the end of the 19th century, Dutch colonial policy shifted from indirect rule to attempts to fully subject the outer islands, especially Sumatra (Locher-Scholten, 1996: 142). In several military operations, the Dutch started to expand direct control and conquered most of Sumatra at the beginning of the 20th century (Locher-Scholten, 1996: 95). The first agrarian regulation, the *Agrarische Reglement*, was enacted to clarify land use and forest management in Jambi and other parts of Sumatra. The legislation provided the legal basis for dividing forests into three categories: permanent forests, forest reserves and forests for the extraction of non-timber and timber forest products (Nurjaya, 2005: 42).

In 1927, the colonial government enacted a new forest law that strengthened the role of the colonial forest service (Galudra and Sirait, 2009: 530). Since this time, the colonial forest service and its post-colonial successors (e.g. Ministry of Forestry, (MoF)) have had the authority to designate land as state forest and to control the use of forest resources (Galudra and Sirait, 2009: 530). Furthermore, forest zoning and the designation of the forest domain was expanded towards Sumatra and other islands (ibid.). However, the colonial authorities lacked the capacities to spatially extend the forest management approaches developed for the colonial core areas to the vast forest areas of Sumatra, Borneo and Papua.

With some exceptions, zoning and forest management on Sumatra existed mainly on paper. Various types of customary law (*adat*)[6] remained the most relevant legal order governing access to forest and agricultural land (Nurjaya, 2005: 42). However, in some areas of Jambi (and beyond), the colonial land and forest authorities induced significant changes, altering pre-existing customary arrangements. For instance, the colonial authorities established protected areas that still exist today. The *Wildreservaat* Berbak (today the Berbak National Park and Berbak Carbon Initiative) was established in 1935, while north of the present-day Bungku village a protected area (today the forest reserve TAHURA Sultan Thaha Syaifuddin) was established in 1933 (Pemerintah Kabupaten Batang Hari, 2010: 8).

In addition, colonial administrative reforms challenged pre-existing administrative structures and customary forms of regulating access and property. In 1903, after the Dutch conquest, the former independent Sultanate of Jambi became part of the Residence of Palembang (today South Sumatra) (Locher-Scholten, 2004: 239). Later, Jambi became an independent Residence but the former jurisdictional division of the territory of the former Sultanate into *Kalbus* was replaced by the administrative system used in Palembang (Guillaud, 1994). From 1919 onwards, the Residence of Jambi received a hierarchical territorial structure divided into *Onderafdeeling* (departments), *Margas* (subdistricts) and *dusuns* (hamlets/villages). *Margas* consisted of five to six villages and were governed by the *Pasirah*. The *Pasirah* was usually a member of the leading lineage in the area and received the authority to legitimate property rights for the non-European population (Galudra et al., 2014: 723; Sevin and Benoît, 1993: 97). The *Pasirah* outlived the colonial regime and remained a relevant authority in property issues in Jambi at least until the New Order regime enacted its village law in 1979. The first land titles issued by the *Pasirah* and the Dutch colonial authorities are still used as a source of legitimacy for stakeholders in land conflicts. The Dutch not only changed the territorial structure, they also introduced rubber and started oil drilling close to the present-day Bungku village. Rubber became one of the most important agricultural commodities of the province during the 20th century.

The development of a colonial forest and land tenure policy on the island of Sumatra at the end of the 19th century had territorial and scalar consequences. Colonial forest authorities established a colonial scale of forest regulation that challenged pre-existing rules for accessing forest resources. Forest zoning, monitoring activities and border demarcations underpinned the territorial claims and the new scale of regulation. Moreover, the colonial forest authorities constructed watershed protection and even climate protection (Galudra and Sirait, 2009: 529) as scales of meaning for legitimizing colonial forest claims. In the Residence of Jambi, the colonial administration introduced a new scale of land tenure regulation and established the *Pasirah* as a new public authority responsible for allocating land to Indonesian citizens (Guillaud, 1994: 125; Sevin and Benoît, 1993: 97).

The early post-colonial period

Indonesian independence induced no abrupt changes in forest and land tenure governance. The Indonesian forest service (*Jawatan Kehutanan*) started activities on Sumatra[7] in 1947 (Barr, 2006: 19; Nurjaya, 2005: 45). The structure and political orientation of the service was heavily influenced by its colonial precursor (Barr, 2006: 19). At least until the 1950s, most of the colonial forest regulations remained valid (Nurjaya, 2005: 46).

In 1957, Government Regulation 64/57 handed far-reaching forest management competencies to provincial governments and provided significant timber revenues for provincial governments (Barr, 2006: 21; Nurjaya, 2005: 46). Christopher Barr (2006: 20) argues that the new regulation represented a political commitment by the government of Indonesia's first president Soekarno to the provincial governments in order to maintain the integrity of the newly independent state. The regulation permitted provinces to issue logging concessions of up to 10,000 ha for up to 20 years. Furthermore, the regulation guaranteed the independence of the provincial forest services (ibid.). The regulation induced the down-scaling of forest management competencies but failed to clearly define the forest estate on the outer islands (ibid. 21). Consequently, the co-existence of customary law and formal law continued for at least the first two decades of independent Indonesia (ibid.).

In 1960, the Basic Agrarian Law (*Undang-Undang Pokok Agraria*) was passed. The BAL still applies (at least for non-forest land) and is highly relevant for understanding contemporary land conflicts and especially agrarian reform movements such as the Indonesian Peasant Union (Serikat Petani Indonesia (SPI)) (Rachman, 2011). The law aimed to harmonize customary law (*adat*) and formal law and replace colonial laws and regulations such as the *Agrarische Wet* (Bachriadi and Wiradi, 2011: 2; Bakker and Moniaga, 2010: 188). The Basic Agrarian Law (BAL) was designed as a holistic law encompassing all natural resources of land, water and air, thus including forest land (Bachriadi and Wiradi, 2011: 1; Presiden Republik Indonesia, 1960b).

The BAL clearly reflects the socialist-oriented zeitgeist of the President Soekarno era. The law stresses the social function of land rights (Article 6), prohibits "excessive ownership" of land (Article 7), postulates the redistribution of land (Article 17 (3)), limits exploitation rights to 35 years (Article 29), guarantees equal opportunities in obtaining land rights (Article 9 (2)) and regulates different types of private property (Articles 16, 20) (Presiden Republik Indonesia, 1960b). The BAL recognizes customary rights (*adat*) as long as they do not contradict the interests of the state (Article 2 (4)), but it does not provide clear regulation on how to solve conflicts between *adat* and formal state law (Barr, 2006: 21). Furthermore, to register and certify rights based on *adat*, the *adat* rights have to be transformed into one of the private property concepts stated in Article 16 (Bakker, 2008: 3). However, according to a study conducted by the University of Palangkaraya and cited by Sandra Moniaga (1993: 139), most of the indigenous communities at the forest margins had no knowledge of the law, were not informed about the necessity to formalize their *adat*-based land claims, and consequently do not hold registered land titles (ibid.).

Broadly speaking, there are today two common lines of criticism of the BAL (Bachriadi and Wiradi, 2011: 4). The first line of criticism argues that the law is dysfunctional because of the lack of implementing regulations, and that it has been used in ways that were not intended by the authors. The second argues that the law has actually increased inequality, since it expands state control and has therefore been used to legitimatize the dispossession of local communities (Bachriadi and Wiradi, 2011: 4).

On the one hand, the wording of the law indicates a strong commitment to policies that promote social and agrarian justice (Bachriadi and Wiradi, 2011: 2). In an interview, one of the leading experts on the Indonesian agrarian movement argued that the law has "[. . .] an inclusive spirit, was socially balanced and gender sensitive and an attempt to abolish large land holdings".[8] The redistribution of land (e.g. land reform) as stipulated by the law was not implemented except for small-scale pilot schemes on Java (Rachman, 2011: 40). On the other hand, the law privileges Western forms of private property over community property, aims to transform collective rights based on *adat* to individual property rights, and strengthens the right of the state to control land (Bachriadi and Wiradi, 2011: 2–3; Bakker and Moniaga, 2010: 188). The New Order regime limited the applicability of the BAL to non-forest land (Barr, 2006: 23). Therefore, today the BAL is relatively weak in legal terms but is still used by NGOs, social and environmental justice movements, and peasant organizations to underpin political campaigns for a more equal distribution of agrarian resources (Bakker and Moniaga, 2010: 88; Hein and Faust, 2014: 25; Rachman, 2011: 54). For example, members of SPI refer to the BAL to legitimize land occupations within the Harapan Rainforest project (see Chapter 5 for more detail).[9]

The BAL can also be understood as an attempt to construct a new national scale for governing natural resources and as an attempt to deconstruct pre-existing colonial scales. For instance, Craig C. Thorburn (2004: 36) stresses that the authors of the BAL "[. . .] envisioned an entire national community guided by an overarching sense of social function. The state, as the ultimate arbiter of 'national *adat*', was in effect granted *beschikkingsrecht* [rights of disposal, translation added by the author] to all the land, sea and natural and economic resources in the country". The post-colonial state sought to legitimize the BAL as a new national scale of regulation by constructing a complementary scale of meaning based on nationalism and on the key role of land for achieving welfare and social justice in rural areas.

New Order

The New Order regime refers to the period under former President Suharto. The autocratic and modernization-oriented regime changed forest and agrarian politics significantly (Bachriadi and Wiradi, 2011: 6). The regime established and stabilized political control over the outer islands, liberalized forest exploitation, standardized and Javanized village administration, expanded the transmigration program and stopped the land redistribution policies initiated by Soekarno (Bachriadi and Wiradi, 2011; Barr, 2006; Kato, 1989; Levang and Sevin, 1990). These political processes changed scales of meaning and regulation (e.g. centralization of forest governance), induced far-reaching landscape transformation processes (e.g. the island of Sumatra lost 25–30 percent of its forest cover, Barr, 2006: 28) and significantly altered the ability of local actors to access forest land.

In 1967, President Suharto enacted the Basic Forest Law. According to Barr (2006: 23), the law provides the first comprehensive legal and administrative framework for managing Indonesia's forest estate. Through the enactment of the Basic Forest Law, Suharto's regime reestablished the dualistic structure of the colonial system – with two separate laws, one regulating forest management and access to the state forest and one regulating access and property outside of the state forest. First, the law defines state forest (*Kawasan Hutan*) as a forested territory or a non-forested territory designated for reforestation (Article 4). It gives the Directorate General of Forestry within the Ministry of Agriculture (later upgraded to the Ministry of Forestry (MoF)) the authority to designate approximately 70 percent of Indonesia's land mass as state forest (Article 7) (Presiden Republik Indonesia, 1967). Second, the law delegates authority to conduct forest zoning to the Directorate General of Forestry and introduces four different forest categories: production forest, recreation forest, forest reserve and nature conservation forest. Third, the law provides the legal framework for commercial forest exploitation and for the economic liberalization of forest management (*Hak Pengusahaan Hutan, HPH*) (Article 14). Fourth, the law states that customary forest is part of the state forest and that the activities of customary (*adat*) communities that contradict the law (e.g. shifting cultivation) are prohibited (Article 17). Furthermore, only the still existing *adat* communities have rights to the forest, with the important addition that the state decides which community is eligible and which is not. Consequently, many local and indigenous communities, including those living within the forests that became the Harapan Rainforest and the Berbak Carbon Initiative more than 50 years later, lost at least de jure access to their customary land.

Forest management during the New Order era aimed to achieve at least three major objectives. First, revenues from forest exploitation were necessary in order to stabilize the state and its various apparatuses (Barr, 2006: 23). This was mainly achieved by establishing clientelistic networks creating mutual dependencies of actors of the civil and military bureaucracy from the central government down to the village level (Barr, 2006: 24; Barr, 1998: 4). Second, revenues from forest exploitation were necessary to stabilize the state budget and enhance the trade balance. Third, forest resources were used to attract foreign investments (Barr, 2006: 27).

In the 1970s, the central government established full control over Indonesia's forest resources and allocation procedures for all types of forest concessions. The central government revoked any rights of the sub-national governments to issue forest exploitation permits (Barr, 2006: 25) and started to allocate concessions to timber companies. The first logging concessions in the area of today's Harapan Rainforest project (e.g. PT Asialog[10] and PT Tanjung ASA)[11] date back to this period. Through up-scaling the permit procedures for all concession types, the central government sought to prevent sub-national elites and their patronage networks from accessing forest resources and to facilitate access to forest resources for the Jakarta-based elite (Barr, 2006: 25–26). Up-scaling finally consolidated Jakarta's attempts to control the forest resources of the outer islands.

Javanizing the outer islands: village law and transmigration

After the central government consolidated control over the forests, another major state project was initiated: Javanizing the outer islands. The Village Law of 1979 (*Undang-Undang tentang Pemerintahan Desa 5/1979*) aimed to homogenize the administrative structure of the archipelago, and the transmigration program was set to promote modernization and development and export Javanese culture to the outer islands (Burkard, 2002: 5; Kato, 1989: 91–94; Warren, 1990: 1–2).

The Village Law was intended to replace the pre-existing structures that were based on colonial concepts or customary law (*adat*) by imposing the Javanese village (*Desa*) concept (ibid.). Indonesia's local governance and administrative structure remained highly diverse until the end of the 1970s. As the lowest administrative unit, villages had different names, meanings and sizes and were led by local leaders with different names, sources of legitimacy, and responsibilities concerning the customary regulation of land and forest tenure (Kato, 1989: 91). This diversity was also acknowledged by Indonesia's constitution of 1945 (Kato, 1989: 114). In Jambi, for example, the colonial system of village administration remained relevant until the enactment of the Village Law in 1979 (Galudra et al., 2014: 723).

The new Javanized village concept introduced a new hierarchical socio-spatial organization that undermined previous forms of socio-spatial organization. According to the law, the village (*desa*) as a jurisdiction should be led by a village head (*Kepala Desa*) as the executive body and by a village council (*Lembaga Musyawarah Desa*) as the legislative body of the village government. The Village Law stipulated the sub-division of the village territory into hamlets (*dusuns*) led by a *Kepala Dusun*. Hamlets are further divided into different neighborhood units[12] (*Rukun Tetangga*) led by the *Ketua RT* (Bebbington et al., 2004: 192; Kato, 1989: 94; Warren, 1990: 3).

Through standardizing and formalizing village government and village administration, the Village Law undermined the traditional customary authorities and the traditional income sources of villages (Kato, 1989: 105). The village territory was imposed as a new scale of regulation contradicting previous scales of regulation, such as the *Wilayah Adat*[13] (customary land) of the Batin Sembilan in Jambi. New village boundaries disrupted previous forms of the Batin Sembilan's socio-spatial organization, which were based on lineages and watersheds.

Furthermore, the Village Law decoupled traditional local leaders from their material basis of power and authority (Bebbington et al., 2004: 193). Their previous sources of power, in Steven Lukes' words their sources of visible power, were based on the right to allocate land and fishing rights and to collect rubber taxes (Galudra et al., 2014: 725; Kato, 1989: 108). In the Berbak area, for instance, land was controlled by the *Pasirah* of the *Marga* of Berbak until the early 1980s.[14] Today, the village head is de facto in charge of land allocation and drainage permits.[15] Galudra and colleagues (2014) identified similar developments in other parts of Jambi.

The aim of the Village Law was not only to standardize scales of regulation. The law was also part of a set of policies aiming to establish national modernization and development as new scales of meaning to replace previous scales of meaning based on local *adat* (e.g. customary authority and locally relevant cultural meanings of landscapes and territories). The New Order regime regarded village diversity and non-Javanese forms of socio-spatial organization as potential threats for development and as signs of underdevelopment (Bebbington et al., 2004: 192; Hoey, 2003: 112; Kato, 1989: 93). As the lowest administrative unit, village governments were to act in line with national development targets. At the same time, they were the lowest part of the national surveillance and control network of the New Order regime (Kato, 1989: 107–113). However, the New Order regime was not fully able to defeat customary authority. Many village governments were not fully operational and in many peripheral villages *adat* and traditional public authorities remained influential, rapidly regaining importance during the *Reformasi*[16] era (Bebbington et al., 2004: 193).

Whereas the Village Law focused on Javanizing the administrative structure, the transmigration program focused on Javanizing the population, land use practices and property relations. Transmigration is a government-led resettlement program and rural development initiative. The program received significant financial support from the World Bank (Kebschull, 1986: 152–153).[17] In many parts of the Archipelago, the transmigration program contributed significantly to the expansion of a Western and individualized concept of property, of cash crops, of wet rice production and of a modern bureaucracy (Armitage, 2002: 211; Cramb et al., 2009; Kebschull, 1986: 37; Li, 2005: 14–15; Roth, 2009: 202).

Transmigration dates back to the colonial period. The *Kolonosatie* program resettled Javanese farmers mainly to Lampung but also to Jambi (Levang and Sevin, 1990: 1,3). The official rational of the transmigration program and of its colonial precursor was to reduce population pressure and land scarcity on Java and Maduro and to redistribute the population to the sparsely populated outer islands (ibid.). Resettlement was also driven by the imagination of the superiority of Javanese and Balinese culture and land use practices (Fearnside, 1997: 559; Levang and Sevin, 1990: 4). Javanese and Balinese transmigrants were framed as model farmers that would persuade "backward" slash-and-burn farmers of the outer islands to use more modern land use techniques (ibid.).

The transmigration program had its peak in the 1980s. By 1989, five million people had been resettled to the outer islands (Fearnside, 1997: 554). Numbers declined in the 1990s but remained substantial (e.g. 90,762 families in 1996–1997) (Potter, 2012: 272). In the *Reformasi* period, the relevance of the project declined significantly (e.g. 2265 families in 2000) (Potter, 2012: 272). Authority over the program was transferred to district governments. Today, potential sending districts and receiving districts directly negotiate with each other (Potter, 2012: 273) and candidates can select between different destinations using an online platform (Dinas Tenaga Kerja dan Transmigrasi Sumatera Barat, 2015; Hein et al., 2018a).

The transmigration program allocated between 1.75 ha and 3.5 ha land to each household for crop production, including official land titles issued by the National Land Agency (*Badan Pertanahan Nasional*, BPN) (The World Bank, 1979: 33, 73). In addition, the transmigration authorities provided a 0.25 ha house lot, seeds and start-up funding (Fearnside, 1997: 555). In Jambi, the first post-colonial projects were developed in the peat swamps of the Batang Hari delta, north-east of today's Berbak Carbon Initiative (Levang and Sevin, 1990: 6). The projects induced enormous landscape transformation processes caused by drainage and land reclamation activities. Additional projects were developed in Jambi's hinterland, for instance in the Sungai Bahar area north of the Harapan Rainforest project.

In many parts of Indonesia, the transmigration settlements overlapped with the customary land used by local communities. In Jambi, transmigration and local resettlement schemes very likely contributed to the transformation of lineage-based property concepts to individualistic and commodified property (c.f. Hauser-Schäublin and Steinebach, 2014; Krishna et al., 2014). A member of the village parliament in Tanjung Lebar, for instance, stated: "In the times before the transmigration project was implemented we did not use land titles in the village [. . .]".[18]

The transmigration program was also used as inspiration and as a source of legitimacy for settlement and forest conversion initiatives by village and district governments that violated the Forest Law and re-claimed former customary land within the state forest territory. The name of an informal settlement within Harapan Rainforest, Transswakarsa Mandiri, makes direct reference to a sub-program of the state-based formal transmigration program. Within the Berbak Carbon Initiative at least two new district-to-district transmigration settlements were recently established, causing conflicts among different state agencies and settlers (see Chapter 5 for more detail).[19]

Reformasi and post-reformasi: rescaling through democratization and decentralization

On 21 May 1998, President Suharto announced his resignation after two years of student protests calling for democracy, and after ethnic violence and sharp economic decline (Hofman and Kaiser, 2002: 3). His successors President Habibie, President Wahid and President Megawati implemented far-reaching democratic reforms and decentralization processes that significantly transformed

Indonesia's political landscape. Indonesia's "big bang decentralization" was nei-ther a controlled nor a planned process (Hofman and Kaiser, 2002). The first decentralization laws and regulations, e.g. Law 22/1999 on Regional Govern-ance, were formulated in an over-hastily way, reflecting the weak power base of the central state during the political transition period after the fall of Suharto (Barr et al., 2006: 2; Hofman and Kaiser, 2002: 3). Decentralization and regional autonomy and new revenue-sharing arrangements among the central govern-ment, provincial and district governments did not follow a stringent plan but were rather an ad hoc response to separatist tendencies, with the aim of main-taining the integrity of the nation state (McCarthy, 2007: 96).

Regional governments including village heads and customary leaders took advantage of the confusion and the political vacuum caused by the weak cen-tral state. They interpreted reforms to their advantage and started to assert far-reaching administrative authority over forests, land tenure and natural resources (Barr et al., 2006: 2; Hein et al., 2016). On the local level, decentralization increased competition between different actors over the new profit options. In some regions of the archipelago (e.g. Central Sulawesi and the Maluku Islands) violent conflicts about access and control of natural resources and political power emerged along ethnic and religious lines (Acciaioli, 2001: 87; Rhee, 2009: 46). Ethnicity and customary law reemerged as means of controlling land and natural resources (Barr et al., 2006: 12; Moeliono and Dermawan, 2006: 109; Rhee, 2009: 109). The reemergence of ethnicity and customary law are highly relevant for understanding contemporary conflicts about access and property in Jambi and other parts of Indonesia (Beckert et al., 2014; Hein and Faust, 2014; Hein et al., 2016; Steinebach, 2013b).

Reforms of forest, land and village governance induced a wide range of changes. However, in many cases and especially in the forest sector, the reforms only temporarily shifted de jure competencies. Most relevant competen-cies remained on the national level. However, the changed political context extended the agency of local public authorities including village governments, customary communities and agrarian reform movements (Barr et al., 2006: 11; Moeliono and Dermawan, 2006: 109; Peluso et al., 2008: 388). Whereas de jure authority over the state forest only temporarily shifted, especially the district governments benefitted substantially from new forest revenue-sharing arrange-ments. Districts in which forest concessions are located now receive a four times higher share of the concession license fee and receive 40 percent of the reforestation fund (Resosudarmo et al., 2006: 61, 67).[20]

Decentralization of the forest sector was initiated through the enactment of Regulation 62/1998 on the Delegation of Partial Authority on the Forest Sector to Regions and Regulation 6/1999 on Regional Governance (Reso-sudarmo et al., 2006: 88). These two regulations and a couple of subsequent decrees issued by the MoF permitted district governments to issue small-scale logging and forest conversion concessions in areas designated as conversion and production forest (Resosudarmo et al., 2006: 99; Indrarto et al., 2012: 27). Fur-ther legitimacy was provided through Law 22/1999 on Regional Governance.

The law delegated authority over various governance functions, such as health, education, agriculture, environmental protection and (to some extent) over agricultural land, to district governments (Barr et al., 2006: 11; Indrarto et al., 2012: 27).

In Jambi, district governments issued Community Timber Extraction Permits (*Ijin Pemungutan Kayu Rakyat, IPKR*) within production and conversion forest, inducing a short logging boom on forest land that became part of the Berbak Carbon Initiative after 2008. In 2002, the central government withdrew the authority to issue small-scale logging and forest conversion concessions from district governments after lobbying by the Association of Indonesian Forest Concession holders, and reestablished full authority over the state forest estate (Resosudarmo et al., 2006: 90). With Regulation 34/2002, the central government abrogated small-scale logging but gave provincial and district governments at least the opportunity to provide recommendations during the permit process for forest concessions (Indrarto et al., 2012: 28; McCarthy et al., 2006: 45; Resosudarmo et al., 2006: 90, 104). These seesaw changes of competencies led to uncertainty and confusion, providing room for monetary and political rent-seeking behavior among actors (Moeliono and Dermawan, 2006: 106; and for similar practices in sub-Saharan Africa, please refer to Lund, 2008: 152).

Today, some district governments continue to issue small-scale forest conversion permits, according to an MoF expert in Jakarta. During the interview, the expert explained: "[. . .] the authority was given to the local government actually with a hope to speed up the process, but the problem then the issued licenses were not used for timber but mostly for estate crop plantations. That's why the decision has been revoked, but they continue to release the license but mostly not for small-scale timber [. . .] the local government mostly related to the issuance for oil palm plantation, small scale although this is not legal".[21]

Forest Law 41/1999 mostly resumes the forest policy of the New Order era, e.g. the law declares that customary forests are part of the state forest and allows access to forests only for formally recognized *adat* communities (Bedner and Van Huis, 2008: 184). However, for instance in its Article 68 on community participation, the law goes beyond New Order policies and provides room for implementing regulations on village forestry (*hutan desa*, Ministerial Regulation 49/2008), community forestry (*hutan kemasyarakatan*, Ministerial Regulation 37/2007) and smallholder forestry (*hutan tanaman rakyat*, Ministerial Regulation 55/2011).

On the one hand, district and provincial governments have de jure only few management competencies today, e.g. management of forest reserves (*Tanaman Hutan Raya* also known as TAHURA) and conservation forests (*Hutan Lindung*) and they have neither the authority to issue concessions nor to decide on forest classification and designation. On the other hand, they have gained significant influence outside the forest sector. This includes spatial planning, environmental impact assessments and the lucrative approval of agricultural plantation and mining permits on non-forest land (Indrarto et al., 2012: 31; Paoli et al., 2013: 27–28).

Like the forest sector reforms, the reform of village administration was characterized by seesaw changes. Law 22/1999 on Regional Governance no longer describes villages as the lowest administrative unit but as "autonomous units", recognizing local customary law and regional characteristics (Moeliono and Dermawan, 2006: 115). Only a few years later, the revised decentralization law of 2004 and the regulation on village government (Government Regulation 76/2001) again stipulated that villages across the archipelago should have a uniform structure (ibid.). Despite this drawback, the reforms have democratized village governments. Village heads and village parliaments (*Badan Permusyawaratan Desa*, BPD) are elected for six years for a maximum of two terms. BPDs and village heads can enact village regulations, prepare village development plans and must be involved in spatial planning processes conducted by higher authorities (Bedner and Van Huis, 2008: 174).

The new laws and regulations have not changed the spatial organization of village governance. Villages are still sub-divided into hamlets and neighborhoods. Although the reforms provided more room for self-determination for village governments, they offer no de jure control over forest resources within the state forest. In any case, de facto former village heads in Bungku and Tanjung Lebar have exercised authority over state forest since the fall of Suharto. Village heads in both villages legitimize their control over the state forest (e.g. demonstrated through the issuance of village-scale land titles and the support of forest conversion and settlement within the state forest) with reference to pre-existing customary territories. These activities are at least partially backed by the older Governmental Regulation 24/1997 on land registration (Article 24) issued under the Suharto Presidency. This regulation gives village heads a formal role in registering *adat* land rights and other rights based on oral history (Nurhaniah, 2006: 74; Presiden Republik Indonesia, 1997).

Access to different types of de jure land and forest rights

Indonesia's forest and land tenure regime is complex and governed through a wide set of laws and regulations (Brockhaus et al., 2012: 32). I have introduced the most relevant ones, e.g. the Forest Law 41/1999 and the Basic Agrarian Law 5/1960. These laws and regulations translate into different types of de jure land and forest rights accessible by different actors, including rights that permit ecosystem service trade (relevant for selling forest carbon credits).

First, it is relevant where the land is located. If land is located within the state forest, it is subject to the Forest Law and related regulations. State forest land is under the authority of the MoF. If land is located outside the state forest, also called areas for other use (APL, *Areal Penggunaan Lain*), it is subject to the Basic Agrarian Law. APL is under the authority of district governments and the National Land Agency (BPN). Second, both laws distinguish between land and forest rights for individuals (smallholders), communities (including formal villages), and corporate actors. Third, the forest regime permits only temporary use rights (*hak pengusaha hutan*) for all actor groups. In contrast, the agricultural regime permits inheritable and alienable individual rights (*hak milik*) for small-holders with a maximum size of 20ha,[22] communal land rights (*hak ulayat*) for *adat* communities and temporary cultivation rights (*hak gunah usaha*) for larger corporate actors (Bedner and Van Huis, 2008: 179–180; Presiden Republik Indonesia, 1960b, 1960a, 1999). However, as we will see later, the de facto rights regime is far more complex than the formal legislations allow for.

De jure rights to forest and land

The allocation of forest concessions is a complex process which has often been described as non-transparent and corruption prone (Casson and Obidzinski, 2002; Smith et al., 2003). To obtain most of the various forest concessions available, actors have to establish a company or a cooperative and have to apply for a permit from the MoF. The legal procedure for accessing ecosystem restoration concessions, for instance, has been described as long and complex (Walsh et al., 2012a; Walsh et al., 2012b). The Forest Law provides different forest concessions for corporate actors (e.g. ecosystem restoration, forest plantation and logging concessions), for village communities (*hutan desa*/village forest concession), for communities (*hutan kemasyarakatan*/community forest concession) and for individual smallholders (*hutan tanaman rakyat*/smallholder forest concession). However, it is usually difficult for village governments, communities and individual smallholders to access these concessions. Smallholder forest concessions (*hutan tanaman rakyat, HTR*) are often allocated by forest authorities as a means to solve conflicts between the forest authorities and smallholders occupying land designated as state forest.[23] This was the case in a village in the southeastern part of the Berbak Carbon Initiative, and it was discussed as a conflict solution tool for solving conflicts between smallholders in Bungku and the conservation company PT REKI, who manage the Harapan Rainforest.[24] In other cases, in Jambi, the allocation of HTR concessions was often facilitated by NGOs or donors. For instance, the NGO Amphal supported by the Finnish Development Cooperation runs two smallholder agroforestry projects based on HTR concessions in the surroundings of the Bukit Duabelas National Park in Jambi.[25]

Asked about how communities could obtain a village forest concession (*hutan desa*), experts from academia, donor agencies and NGOs interviewed in July 2012 mentioned that requirements such as the preparation of management plans and forest inventories, and the levying of administrative charges represent significant barriers to local communities and smallholders that cannot be resolved without external support.[26] In addition, only state forest that is designated as production forest (*hutan produksi*) and that is not yet allocated is eligible for smallholder forest concessions. Village and community forest concessions represent an exception. They can also operate in conservation forests (*hutan lindung*). Furthermore, the respective regulation on village forests clearly links village forests to existing and formally recognized villages (Menteri Kehutanan, 2008a). This excludes groups that settle informally or live nomadically inside the state forest.[27] The indigenous rights movement AMAN rejects community and village forest concessions and argues "[. . .] we do not agree with that, because actually the basis of these forest categories is wrong. Because the forestry law itself is wrong [. . .] it doesn't recognize indigenous people's rights; [. . .] the law is wrong and consequently implementing regulation is also wrong".[28] Community and village forest concessions remain part of the state forest territory, thus they remain under the authority of the MoF. They only permit temporary use rights. AMAN consequently refuses concessions and demands full land ownership and the recognition of indigenous territories.[29]

The recent constitutional court review declared that *adat* forest is no longer part of the state forest but is subject to the rights of *adat* communities (Rachman, 2013: 2). In 2015, after the election of President Jokowi, plans were announced to facilitate and expand the allocation of community and smallholder forest concessions and to implement the decision of the constitutional court (HuMa, 2015; Kompas.com, 2016; Nugraha et al., 2017).

Furthermore, in specific circumstances communities with the support of the district government can apply for an enclave. An enclave implies the release of a specific area from the state forest. The MoF then assesses the request according to specific criteria covering inter alia ethnicity, the history of the area, evidence of community rights and ecological indicators.

While a number of different corporate concessions have existed (e.g. logging concessions, forest plantation concessions) since 2008, a new concession type has been established that permits private actors to conduct conservation and ecosystem restoration activities (see Chapter 4 and Walsh et al., 2012a and 2012b). Ecosystem restoration concessions are private conservation concessions for habitat protection, habitat restoration and management (Ministry of Forestry, 2008: 5). They permit the commercial exploitation of non-timber forest products and the commercial use of ecosystem services such as ecotourism, biodiversity and carbon capture (ibid. 7). The conservation company PT REKI who runs the Harapan Rainforest project holds a concession of this sort.

In the process to obtain land rights in areas designated as APL, district governments and the National Land Agency (*Badan Pertanahan Nasional*, BPN) have the final voice. In peripheral areas without land registration, the village heads are (according to Governmental Regulation 24/1997 on land registration) in charge of land deeds and issuing the documents that are required to convert customary ownership to a formal land title (*hak milik*).

The BPN distinguishes between two procedures for registering land as *hak milik*. The first, called *sistematik*, refers to government-sponsored and government-assisted land-titling schemes conducted in specific areas, such as the *Proyek Operasi Nasional Agraria* (PRONA), for agricultural land. In this case, village heads and civil servants from the BPN jointly conduct tenure mappings to identify and certify land. The second procedure is called *sporadik* and refers to the process of individual land registration. The certification of a plot of land requires several documents, including a declaration confirming that land taxes have been paid, a document on the ownership history, a document that describes the location of the land and its borders (*Surat Pernyataan Penguasan*

Fisik Bidang Tanah), and a declaration that the plot is not claimed by another party. To convert customary rights, a statement from the village head is necessary. However, in many parts of Jambi, the term *sporadik* customarily refers to a village-scale land title, which is even recognized in part by the BPN in Jambi. According to a BPN officer in Jambi, a *sporadik* is not a legal land title but it can be considered as a *Surat Pernyataan Penguasan Fisik Bidang Tanah* and can be accepted by the BPN as a document legitimizing ownership by the village government.[30] The document is needed for a land title application for previously untitled land (Presiden Republik Indonesia, 1997; Wibisono, 2012: 9–11).

To obtain a concession for cultivation rights (*hak gunah usaha* (HGU)) corporate actors have first to apply for a location permit (*ijin lokasi*) from the district government. Before issuing a location permit, the district government has to consult local communities potentially affected by the concession (Paoli et al., 2013: 24).[31] After holding the location permit, the planned plantation estate is subject to an environmental impact assessment involving the environmental protection authority of the district (ibid. 28). In a last step, the BPN issues the HGU (ibid.). In some cases, e.g. in Seponjen, companies run plantation estates based only on the location permit without having conducted an environmental impact assessment and without having a concession for cultivation rights (HGU).[32]

The different types of land and forest rights described above are subject to different laws and regulations and involve a range of authorities such as village governments, district governments, the BPN and the MoF/MoEF. Each authority has its specific scale of regulation which reflects the position and the different dimensions of power of authority within the different apparatuses of the state. However, as the previous sections have shown, power constellations and scales of regulation are not stable but are continuously re-negotiated. Table 3.1 shows an overview of the different rights or, in other words, property types, their legal basis, limitations, the authorities involved and lists of stakeholders eligible for each property type.

De jure land rights and state-based settlement schemes

As outlined above, transmigration settlements were established in many parts of Jambi during the New Order era. The program participants usually also received titled agricultural land. The settlements in the Sungai Bahar area north of the Harapan Rainforest (developed between 1983 and 1986) provided approximately 2 ha of agricultural land and a 0.25-ha plot for a house, yard and garden. However, a member of the village government argued that the local population had difficulty joining the program and receiving the pledged land. He reported that some Batin Sembilan households paid a fee to join the program but received no land from the transmigration authorities.[34] A Batin Sembilan household head said that a total of 56 Batin Sembilan households from Tanjung Lebar joined and received start-up credits and land titles issued by the National Land Agency.[35] Some of them later sold their land to migrants.

Table 3.1 De jure land and property rights

Property type	Regulation/law	Limitations	Authority involved	Eligible stakeholder	Eligible land and forest category
Collective customary rights (Hak Ulayat)	Basic Agrarian Law 5/1960	Communal land title based on customary rights	District government, BPN	Customary communities	APL
Individual land title (Hak Milik)	Basic Agrarian Law 5/1960, Governmental Regulation 24/1997 and 56/1960	Agricultural land with a max. size of 20 ha	Village head and BPN	Individual land users	APL
Cultivation permit (Hak Guna Usaha, HGU)	Basic Agrarian Law 5/1960, Governmental Regulation 26/2007	Concession for agricultural plantations (e.g. oil palm), for max 60 years	District government, BPN	Corporate actors	APL
Smallholder forest concession (Hutan Tanaman Rakyat, HTR)	Ministerial Regulation 55/2011	Max land size of 15 ha per household, for max 60 years	Ministry of Forestry	Individual land users	Permanent Production Forest
Community forest concession (Hutan kemasyarakatan)	Ministerial Regulation 88/2014	Concession for community forestry, for max 35 years	Ministry of Forestry	Community with community association	Production and Conservation forest
Village forest concession (Hutan Desa)	Ministerial Regulation 49/2008	Concessions for village forestry, for max 35 years	Ministry of Forestry	Village community	Production and Conservation forest
Customary forest (Hutan Adat)	Forest Law 41/1999 and Putusan MK No. 35/PUU-X/2012 tentang Hutan Adat	Implementing regulation pending	Ministry of Forestry and Constitutional Court	Customary communities	n. a.
Forest plantation concession (Hutan Tanaman Industri, HTI)	Forest Law 41/1999, Government Regulation 7/1990	Concession for timber plantations, max 35 years	Ministry of Forestry	Corporate actors	Permanent Production Forest
Logging concession (Pemanfaatan Hasil Hutan Kayu)	Forest Law 41/1999	Logging concession, max 20 years	Ministry of Forestry	Corporate actors	(Limited) Production Forest
Ecosystem restoration concession/conservation concession (Restorasi Ekosistem)	Ministerial Regulation No. P. 61/Menhut-II/2008	Concession for ecosystem restoration, for max 95 years	Ministry of Forestry, provincial government	Corporate actors	Production Forest
Demand for enclave/release from forest land	Forest Law 41/1999	Release from state forest (reclassification)	Ministry of Forestry, provincial and district government	Local communities	All types

(Sources: cited laws and regulations[33])

The program for underdeveloped villages (*Impress Desa Tertinggal*) and the *Transsos* settlement program were presented as social programs to "develop the local population". They were developed from the 1970s until the late 1990s. The aim was to sendentarize the semi-nomadic Batin Sembilan families; the programs provided them with standardized wooden houses. The program in Bungku provided no agricultural land; in Tanjung Lebar a few participating households received 1 ha of agricultural land (Table 3.2).[36]

Jambi's contested landscapes: from dispossession and development to conservation

Today, large parts of Jambi are covered by oil palm, rubber, and pulp and paper plantations. Jambi's economy is dominated by agribusiness and by the exploitation of black coal, oil and gas. Because of agricultural expansion and open pit mining, Jambi lost 76,522 ha of forest per year between 2006 and 2009 (Perbatakusuma et al., 2012: 55). Jambi's high deforestation rate and the conversion of peat swamps lead to net annual GHG emissions of 57 MtCO2e (Pemerintah Provinsi Jambi, 2012).[37] The area under oil palm cultivation in Jambi has increased from 44,000 ha in 1990 to 714,399 ha in 2016 (Direktorat Jenderal Perkebunan, 2017). This rapid expansion of the oil palm frontier continues to be a highly conflictive and contradictive process, in particular because conservation areas have also been expanded significantly. Corporate and smallholder oil palm plantations as well as the new conservation areas challenge pre-existing land-use practices such as gardening, swiddens, logging and hunting and gathering, and the related customary regulations (Hein et al., 2016).

The landscapes surrounding the Harapan Rainforest project and the Berbak Carbon Initiative have different histories. I argue that unraveling them helps to improve understanding of the conflict dynamics in the two areas. Both are impacted by an unprecedented oil palm boom but there are differences in landscape transformation and especially in the agents driving the change and the conflicts. While the area of the Harapan Rainforest has been transformed by periods of violent primitive accumulation, the transformation of the peat swamps of the Berbak region was less conflictive and mainly driven by Bugi seafarers and Malay elites.

Landscapes of the forest of hope: dispossession, conflict and interethnic alliances

In pre-colonial times, the landscapes around todays Harapan Rainforest were the home of semi-nomadic Batin and Kubu tribes (Andaya, 2008: 205; Hagen, 1908: 19–20; Hein et al., 2015: 5). Leonard Y. Andaya (2008: 205) describes the Batin of the landscapes between Batang Hari and Musi Rivers "[...] as collectors of forest products for former Malayu Kingdoms, they filled a complementary niche that helped them to maintain a distinctive lifestyle and ethnic identity". According to colonial sources interethnic marriages between Batin and Malay

Table 3.2 Different de jure land allocation and titling schemes for APL land (non–forest land)

Program	Scope	Authority and scale	Land title	Identified in village?
Transmigration	Resettlement from provinces with high population density, plot for home garden and agriculture, housing	Since the *Reformasi*: district government; previously the Ministry for Transmigration	Yes	Kampung Laut and Tanjung Lebar[1]
Program for underdeveloped villages (*Impress Desa Tertinggal*)	Development support for "backward villages"	District government	Not in all cases	Tanjung Lebar
Social housing program for nomadic groups (*Transsos*)	Sedentarization of nomadic and semi-nomadic groups	District government	Only land for house, yard and garden	Bungku, Tanjung Lebar
National Agrarian Operation (*Proyek Operasi Nasional Agraria*, PRONA)	Land-titling program	Ministry for Agriculture and Spatial Planning and National Land Agency	Yes	Kampung Laut (initial survey conducted)
Sporadik	Individual land-title application process	Village government in charge of converting customary claims to land, title issued by National Land Agency	Yes	All study villages

(Source: compiled by the author)

1 Tanjung Lebar was not a transmigration village, but transmigration settlements were developed within the territory the village population claims. Today, those settlements are independent villages.

were common and many Batin increasingly adopted the Jambi Malay language, lifestyle and religion during the colonial period (Andaya, 2008: 205). Today, most Batin Sembilan speak Malay and describe themselves as Muslims.

The Batin Sembilan groups living in the borderland between Jambi and South Sumatra trace their origins back to these groups (Hein et al., 2016: 384; Steinebach, 2013b: 71). The hamlets of Pangkalan Ranjau (Hagen, 1908) and Sungai Beruang founded by Batin Sembilan and today part of the village of Tanjung Lebar, confirm their pre-colonial presence in the area (Figure 3.1 and Table 3.3). The villages of Tanjung Lebar and the neighboring village of Markanding[38] were the oldest settlements in the Harapan landscape, explained a Batin Sembilan elder.[39] The settlements were cultural and socio-economic centers of the Batin Sembilan. Over the year, the Batin Sembilan stayed alternately in the settlements or in the forest shelters to hunt, gather forest products, care for fruit trees and practice shifting cultivation (Hidayat, 2012: 49).

The Batin Sembilan share a common myth of origin. According to this myth they descend from nine sons of a legendary noble named Raden Nagosari. Each son settled on the shore of one of nine rivers (Bulian, Bahar, Jebak, Jangga, Pemusiran, Burung Antu, Telisak, Sekamis, Singoan) in the borderland between the Sultanates of Jambi and Palembang (c.f. Hidayat, 2012: 3).[40] This myth of origin is reflected in the traditional and, at least in part, still relevant socio-spatial organization of the different lineages of the Batin Sembilan.[41] In pre-colonial and colonial times different lineage leaders of the Batin Sembilan controlled forests in the upper watersheds of the Bulian, Bahar and Lalang Rivers (Figure 3.1A). Extended families controlled the forests in a sub-watershed, permitting only family members to establish fruit gardens and dry rice fields.[42] Today, different lineages and extended families of the Batin Sembilan still refer to these and other rivers to explain and legitimize the borders of their ethnic territories (Table 3.3).

The traditional leadership structure of the Batin Sembilan and their socio-spatial organization has been transformed and overlaid by colonial and post-colonial administrative and jurisdictional reforms. In the villages of Bungku and Tanjung Lebar, much of the knowledge of traditional authorities and traditional socio-spatial organization seemed to be lost. The remaining information on the role of traditional authorities and on the hierarchy of authorities is rather contradictive. Rian Hidayat (2012: 71) argues that the *Depati* was the highest authority of the Batin Sembilan during the Sultanate. The Depatia usually governed a *dusun* (hamlet). A Batin Sembilan leader in Tanjung Lebar argued that the *Temenggung*[43] was the highest authority in a *dusun*, positioned above the *Depati*. The *Temenggung* was a cultural and political leader responsible for conflict mediation. Batin Sembilan elders that I met argued in line with Hidayat, or they did not know a *Depati* at all or explained that the *Depati* was called *Mangku* in the settlements around the Harapan Rainforest.

In the early 20th century, the Dutch took control and, as described earlier in this chapter, introduced the administrative system of Palembang and established the *Pasirah* as a new authority governing a *Marga* and holding the authority to

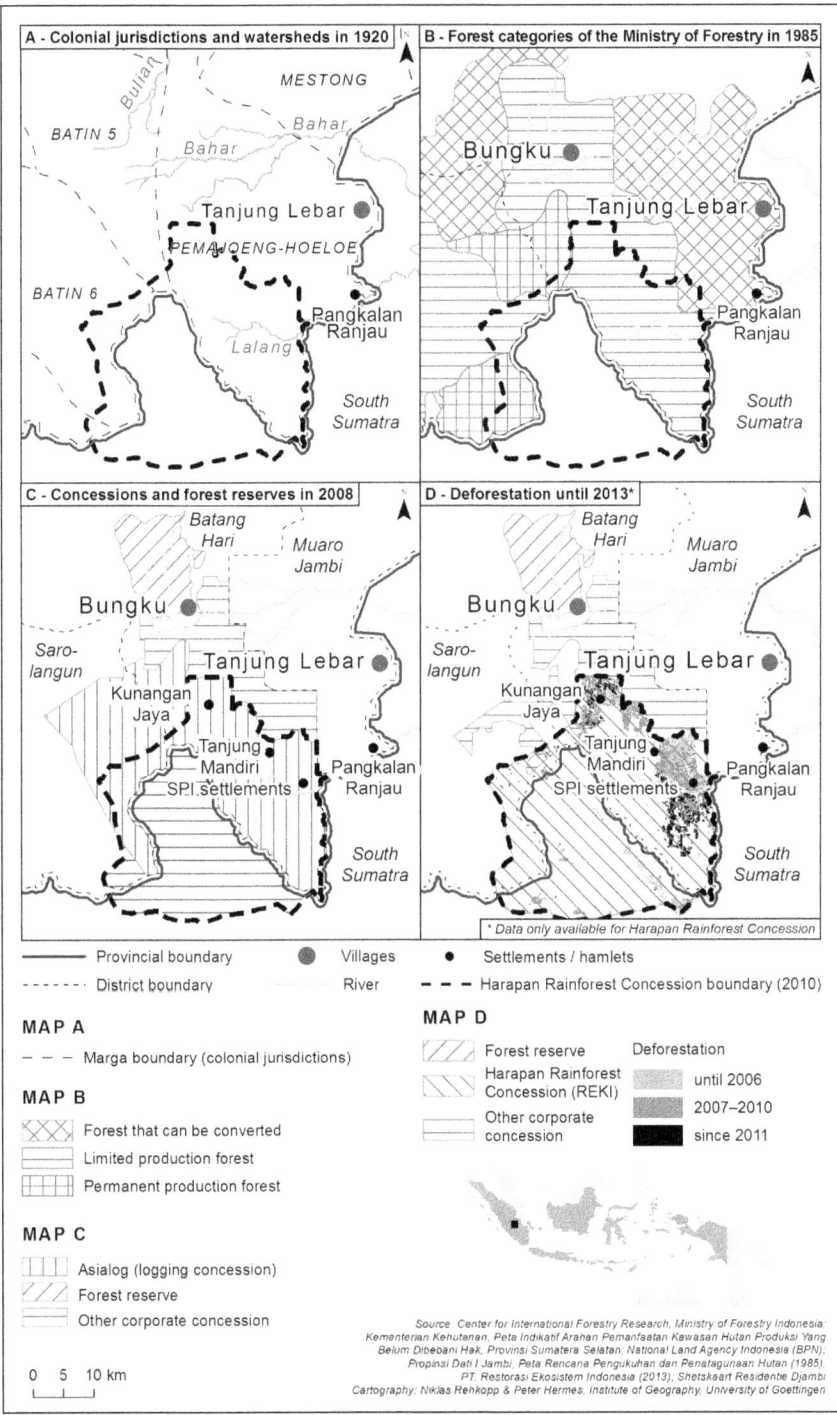

Figure 3.1 Different thematic maps of the Harapan Rainforest and its surroundings
(Source: adapted, based on Hein et al., 2016)

Table 3.3 Lineages of the Batin Sembilan and their settlements

Name of Batin sub-group and river name	Old settlement	Official villages
Bulian (Sungai Bulian)	Dusun Singkawang	Kel. Bulian, Kel. Sridadi, Kel. Pasar, Kilangan, Singkawang, Sungkai, Kel. Bajubang, Petajin, Batin
Bahar (Sungai Bahar)	Dusun Pinang Tinggi, Dusun Padang Salak, Susun Tanah Menang, Dusun Mengkanding, Dusun Tanjung Lebar	Pelembang, Nyogan, **Tanjung Lebar**, Tanjung Pauh Illir, Ladang Peris, Sungai Landai, Dusun Baru, **Bungku** and transmigration villages in the Kec. Sungai Bahar and Bahar Selatan
Other sub-groups: Jebak, Pemusiran, Singoan, Jangga, Burung Antu, Sekamis, Telisak		

(Source: Hidayat, 2012: 33 and own investigations)

allocate land. The landscape became part of the *Marga* of Pemajoeng Hoeloe and the first roads were built linking the region with Jambi city (Tidemann, 1938: 251). The Dutch colonial administration also sought to settle and to "civilize" the native population, challenging their semi-nomadic lifestyles.

After independence, the region initially remained relatively stable. Fundamental transformation processes were triggered by Suharto's autocratic New Order regime (1967–1998). Almost the entire region was classified as state forest area despite the presence of peasant and indigenous communities such as the Batin Sembilan (Figure 3.1B) (Hein et al., 2016: 384–385). To fight "backwardness" and "unproductiveness", in other words to promote "development", the central government allocated large forest areas claimed by the Batin Sembilan to logging companies in the 1970s and to oil palm companies in the 1980s. In the 1970s, the forests in and around today's Harapan Rainforest were divided into three corporate logging concessions. The southern part was assigned to PT Asialog, the northern part to PT Tanjung ASA and the eastern part to PT Suka Rimba Raya (Badan Inventarisasi dan Tata Guna Hutan, n.a.). The classification of the area as state forest and the allocation of state forest to logging companies illegalized the settlements and livelihood practices of the Batin Sembilan and led to de jure dispossession of the Batin Sembilan. De facto, many Batin Sembilan groups were still able to practice shifting cultivation and domestic logging was tolerated, but the companies restricted access to the concession for non-local groups.[44] A Batin Sembilan leader in Bungku argued that the logging company PT Asialog accepted the land rights of the Batin Sembilan.[45] However, some Batin Sembilan had other experiences with the company and reported that shifting cultivation and crop cultivation were not allowed within PT Asialog's concession and early conflicts, especially between Batin Sembilan and the PT Asialog, were also reported, e.g. when staff of the company logged fruit trees belonging to the Batin Sembilan.[46]

Furthermore, as mentioned above, the Law No. 05/1979 on Village Governance imposed the Javanese village administration and jurisdictional system and established villages (*desa*), hamlets (*dusun*), and neighborhoods (*rukun tetangga*) as new scales of regulation (Hein et al., 2016). For instance, the foundation of Bungku village and the establishment of new jurisdictional boundaries between the villages were superimposed on the scales of regulation established by lineages of the Batin Sembilan, and ignored the previous separation of the different lineages living along the rivers of the Bulian River watershed (Batin Bulian) and those living along the rivers of the Bahar and Lalang River watersheds (Batin Bahar) (Hidayat, 2012: 28).

In the 1980s and early 1990s, the allocation of large-scale oil palm concessions and the development of the state-sponsored transmigration resettlement scheme induced the most disruptive transformation processes, which have some similarities with what Karl Marx (1887) has described as primitive accumulation. The vast logging concession still provided the possibility for commercial large-scale resource extraction to coexist with the livelihood patterns of the Batin Sembilan. In the 1980s, a number of transmigration settlements were established in the north of today's Harapan Rainforest project. Through the transmigration program, land was allocated to Javanese farmers acting as out-growers for the state-owned oil palm plantation company PT Perkebunan Nusantara VI (PTPNVI). Only very few Batin Sembilan were able to participate in the program and many of them had difficulties earning a living in the settlements and moved back to their traditional settlements.[47] In 1986, the establishment of the 20,000 ha oil palm estate of PT Bangun Desa Utama (PT BDU), today named PT Asiatic Persada, led to the displacement and dispossession of many Batin Sembilan and to a still-unresolved violent land conflict (c.f. Steinebach, 2013b; Beckert et al., 2014). The new transmigration settlements and the plantation estates led to the fragmentation of the customary land of the Batin Sembilan and especially of the territory claimed by the village elites of Tanjung Lebar. Today, Tanjung Lebar has no contiguous village territory; its five hamlets are scattered and located within and north of the Harapan Rainforest.

Settlement policies for Batin Sembilan groups, such as the program for "underdeveloped villages" (*Impress Desa Tertinggal*), provided no land, or only very limited land, for agriculture. The provision of settlements without land and the eviction of the Batin Sembilan from their customary land that they previously used for swiddens and forest gardens freed up their labor for the demands of the rising plantation sector. Thus, the companies that contributed to the displacement of Batin Sembilan employed many of those that lost access to their forest gardens and swiddens. Others moved south into the forests that later became the Harapan Rainforest, or engaged in hidden resistance strategies such as stealing and selling the fruits of the corporate oil palm plantations.

Following the regime change in 1998, political turmoil, decentralization and democratization induced the most recent and conflictive rescaling processes that have been ongoing in the region (Hein et al., 2016). District heads (*Bupati*), village heads (*Kepala Desa*) and customary leaders took advantage of the political vacuum, interpreting the reforms to their advantage and asserting far-reaching administrative authority over forests (Barr et al., 2006: 2; Hein et al., 2016: 384). Within the areas in and around today's Harapan Rainforest, Batin Sembilan increasingly started to accentuate their social identity as an indigenous group. Hidayat (2012: 3) argues that the active reconstruction of the ethnic identity of the Batin Sembilan should be considered as a process of resistance against the neglect of their rights to land and natural resources, which was especially prevalent under the New Order regime. Furthermore, customary leaders of the Batin Sembilan began using their agency to reestablish their former customary land as a relevant scale of meaning and regulation. Batin Sembilan elites started to collaborate with peasant organizations such as SPI and adopted the land-use practices of mainly Javanese migrants (e.g. oil palm cultivation). They began allocating the forests of the 67,000 ha PT Asialog concession (abandoned since 2003; Figure 3.1C) to arriving migrants (Hein et al., 2016) and started to openly resist the oil palm company PT Asiatic Persada that had displaced them at the end of the 1980s.

Interethnic marriages between Batin Sembilan and migrants gained importance as strategic alliances (Hein et al., 2016: 384; Steinebach 2013b: 74). They enhanced the social status of the indigenous Batin Sembilan and provided migrants with access to land, timber and non-timber forest products within the state forest. The assigning of land to migrants can be considered a socio-spatial resistance strategy of the Batin Sembilan elite to regain authority over state forest land, which they lost during the New Order period, and to reestablish their customary land claims within the state forest (Hein et al., 2016: 384–385). Leaders of the formal village governments – in most cases, Batin Sembilan or individuals with strong kin-ties to Batin Sembilan – legitimated the land transactions by allocating forest conversion permits and village-scale land titles (ibid.). In 2010, the MoF challenged land allocation by village and customary authorities and turned the abandoned logging concession of PT Asialog into a conservation concession, ignoring the existence of settlements and farmland and assigning the area to the conservation company PT REKI (Figure 3.1C, D) (ibid.). The land conflicts emerging in the context of the implementation of the conservation concessions are described in Chapter 5.

Historical and customary rights regime within and around today's Harapan Rainforest

As described above, the Batin Sembilan lived mainly close to rivers in order to be near water, fish and transport. Different lineages and sub-lineages of the Batin Sembilan controlled the adjacent land and forests. The landscape's watersheds can be regarded as scales of meaning (they contain spiritual meanings and are linked to a lineage's specific myth of origin) and as scales of regulation (each lineage has authority over land within its territory). The first (extended) family to convert forest and establish a swidden in a specific area generally owned the land.[48] Other families needed permits to establish swiddens in that area, and land rights continued after the forest had regrown. Batin Sembilan in the hamlet of Kunangan Jaya 1 (part of Bungku village) explain that only members of their lineage are allowed to live in their neighborhood (RT, *rukun tetangga*) and to convert forest along the upper Kandang River (a tributary of the Bahar River).[49] Specific trees, watercourses, hills and other landmarks demarcated the borders of land that was held by various extended families and lineages.[50]

As mentioned above, details of the traditional leadership structure are contradictory, making it difficult to assess the role of traditional Batin Sembilan public authorities in terms of land governance, and especially their role in granting access permits to non-lineage members and other outsiders. A key informant explained that when outsiders stopped using land for more than six months, their rights to the land reverted to the *Temenggung*.[51] The suggestion that the *Temenggung* had authority over land issues is confirmed by a study conducted by the Forest Peoples Program (FPP). Marcus Colchester and colleagues (2011: 10) state that the *Temenggung* "[...] had overall authority over adat territories". Interviewed Batin Sembilan elders generally agreed that outsiders – migrants – needed permits from Batin Sembilan leaders to convert forests.[52] This is also confirmed by colonial sources. Tideman (1938: 78) explains that migrants had to pay rent to Batin Sembilan groups to access land rights. In cases of border disputes or violations, such as the destruction of another family's fruit tree, compensations were negotiated for any losses.[53]

As population density was very low in pre-colonial and colonial times, land was abundantly available. At that time, tree property (fruit trees and *sialang* trees[54] with beehives) might have been more relevant than land rights. Sialang tress with beehives, for example, were owned by specific families and only family members had the right to harvest honey.[55] Fruit trees were usually owned by the families that planted or took care of them. Batin Sembilan planted trees such as durian, langsat and rubber on abandoned swiddens. Old fruit trees in the remaining forest patches of the Harapan Rainforest provide evidence of the historical presence of Batin Sembilan groups. Present-day rubber and fruit tree agroforestry can also be considered as relics of Batin Sembilan forest gardens that existed before the deep transformation processes that began with the New Order.

As described above, Dutch jurisdictional reforms for the Netherlands East Indies changed pre-colonial scales of meaning and regulation (e.g. watershed scale) and lineage-based access and property relations. A Batin Sembilan elder in the Harapan landscape explained, "We were colonized by the Dutch. The Dutch formed a new government and we received a *pasirah* [...] replacing the authorities based on *adat*".[56] In Bungku, the head of a Batin Sembilan family explained that to convert forest, a permit was required from a *pasirah*.[57]

In the late 1980s, the Batin Sembilan's customary regulations in the landscapes around today's Harapan Rainforest project became less important; today, the land claims of different Batin Sembilan families, sub-groups and lineages appear to be contested: their leaders and other public authorities interpret the borders between lineages and extended families in various ways. A Batin Sembilan elder in Tanjung Lebar argued, for instance, that all Batin Sembilan families in Tanjung Lebar, Bungku and Markanding (located west of Tanjung Lebar) have the same ancestors but do not share the same territory.[58] He explained that the territory of the Batin Sembilan groups of Tanjung Lebar is located between the mouth of the Sungai Kandang River and the Mandiangin village water divide, which is located west of the Harapan Rainforest in the neighboring district of Sarolangun.[59] A member of the Tanjung Lebar village government argued that the old border between the Markanding and Tanjung Lebar groups was located in the area of Kunangan Bawah, which is now part of the village of Bungku, and claimed that the whole PT Asialog concession had been part of their territory.[60] A Batin Sembilan elder in Bungku challenged these oral records and maintained that his family's territory was located between Sungai Kandang and Bahar: it was neither part of the land controlled by the Batin groups from Markanding nor Tanjung Lebar.[61] Descendants of the former Batin Sembilan leader, *Depati* Jentikan, claimed that their ancestor controlled the forests around the main hamlet of Bungku, reaching across watersheds to Air Hitam in the district of Sarolangun.[62] Members of the 113 group of Bungku, a local resistance group formed by the Batin Sembilan and Javanese migrants that occupies parts of the oil palm concession of PT Asiatic Persada (for more details on the conflict with the 113 group consider Chapter 5; Steinebach, 2013b; Beckert et al., 2014), argue that Bungku is located in the border area between two lineages (Batin Bulian and Batin Bahar).

The New Order politics of dispossession transformed the Batin Sembilan from landowners to landless peasants, illegal settlers, squatters and day laborers. The rescaling of forest and land governance facilitated resource exploitation by corporate actors but significantly reduced Batin Sembilan groups' abilities to access land and property. Only a few Batin Sembilan families, especially the elites and those who had adopted commercial farming, were able to benefit from the extra leeway provided by the rescaling processes of the *Reformasi* era. Spontaneous in-migration and the transmigration scheme further altered the landscape's demographic structure. Today, Batin Sembilan groups are minorities in Bungku and Tanjung Lebar. According to PT REKI data, only 10 percent of the population in the informal settlement of Transwakarsa Mandiri (part of Bungku) belong to local ethnic groups (REKI, 2011a). A survey conducted by the Forest Agency of the Batang Hari district (2012) in the informal settlement of Bukit Sinjal (part of Tanjung Lebar) shows a similar situation.

Landscapes around Berbak National Park: in-migration, drainage and logging

The Berbak landscape was sparsely populated until the beginning of the 20th century (Sevin and Benoît, 1993: 95). A few small Malay settlements existed on the river banks of the lower Batang Hari River (Claridge, 1994: 290; Hein et al., 2018a; Sevin and Benoît, 1993: 97). The coastal part of the region was home for Orang Laut tribes which were mainly involved in trading, fishing and piracy (Andaya, 2008: 182; Locher-Scholten, 2004: 58). The Malay villages in the delta consisted only of a few houses, and interaction between the villages was limited (Sevin and Benoît, 1993: 97). However, in contrast to the Batin people of Jambi's hinterland, the Malay settlements were under the full authority of the Sultanate (Locher-Scholten, 2004: 48).

The influence of the colonial administration was relatively strong. A first colonial trade and military post dating back to the 17th century (Tidemann, 1938: 337) was located on the banks of the Batang Hari River. The Batang Hari delta west of the Berbak National Park was a target location for the colonial resettlement project *Kolonosatie*, a precursor of the transmigration program of the postcolonial government that started in 1905 (Sevin and Benoît, 1993: 104). As in the area of today's Harapan Rainforest, the Dutch divided the landscape into *Margas* led by a Pasirah (ibid. 97). The area of today's Air Hitam Laut was part of the *Marga* Berbak, Seponjen was part of the *Marga* Kompoe-Hilir and Kampung Laut of the *Marga* Djeboes (Tidemann, 1938: 251). In 1935, Dutch colonial authorities established the Berbak *Wildreservaat* as a precursor of the Berbak National Park (Claridge, 1994: 288). The protected area covered 190,000 ha of peat swamp and mangrove forests (Balai Taman Nasional Berbak, 2013).

In the 20th century, Sumatra's east coast became a popular destination for Banjar from East Kalimantan and for Bugis from South Sulawesi (Benoit and Sevin, 1993: 257; Galudra et al., 2014: 719, 722). Banjar people were experienced hydraulic engineers. In their homelands in East Kalimantan, they converted and drained peat-swamp forest for crop production (Galudra et al., 2014: 722). In Jambi, they mainly settled in the northern parts of the province close to the border with the province of Riau. Bugi immigration intensified during a period of political violence in South Sulawesi in the 1950s and 1960s. The Bugis adopted the drainage techniques of the Banjar, engaged with local political authorities to access land and started peat-swamp conversion and settlement projects along the coast and in the Batang Hari delta (Claridge, 1994: 290; Sevin and Benoît, 1993: 102). More profound changes of landscape and socio-spatial organization started in 1960 and intensified during the New Order era. Transmigrants and the continuing influx of spontaneous migrants from Java and South Sulawesi induced rapid population growth in the Batang Hari delta (Claridge, 1994: 292). The village of Air Hitam Laut located on the margins of the Berbak Carbon Initiative, for instance, was founded by Bugi seafarers[63] from the Luwu district in South Sulawesi in 1965.[64]

The villages of Kampung Laut and Seponjen (located on the western border of the Berbak Carbon Initiative, Figure 3.2) remained relatively stable until the 1990s. The Malay lived mainly from fishing, the gathering of non-timber forest products (especially tapping of *Dyera costulata*, *syn. D. laxiflora* and gathering of rattan), rubber and fruit tree cultivation. The settlements and the fruit tree and rubber gardens were located on the higher alluvial deposits; the lower marshes were covered by peat-swamp forests. Small marsh areas were used for wet rice cultivation. In 1985, the MoF released large state forest areas on both sides of the Batang Hari River and of the Kumpeh River (which is a river

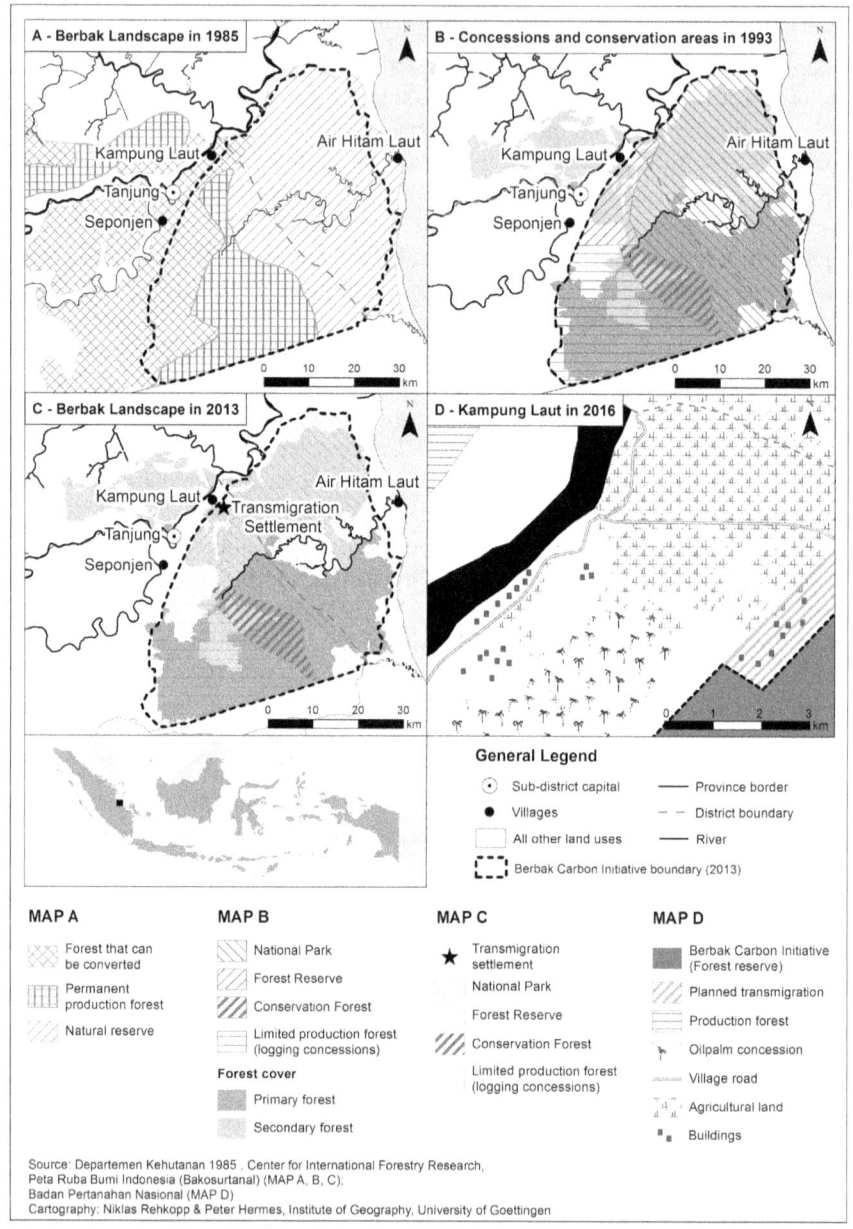

Figure 3.2 Different thematic maps of the Berbak Carbon Initiative and its surroundings
(Source: adapted and based on Hein et al., 2018)

arm of the Batang Hari), permitting the expansion of agricultural production (Badan Inventarisasi dan Tata Guna Hutan, 1985). Yet the exact borders of the reclassified area remained unclear and a 1993 MoF map indicates that at least parts of the released land were again reclassified as state forest (Figure 3.2A, B) despite the ongoing activities of the local population (e.g. logging and gathering of rattan). The forests east of the villages were designated as a forest reserve (TAHURA Sekitar Tanjung) (Badan Inventarisasi dan Tata Guna Hutan, 1993). A few years later, in 1997 (Figure 3.2B), the *Hutan Suaka Alam* (former *Wildreservaat*) became the Berbak National Park, but the size of the protected area was reduced by almost 30,000–162,700 ha (Balai Taman Nasional Berbak, 2013).

In Seponjen, significant village expansion and peat swamp conversion started in the 1990s. In 1992, Pak Hadji Pattimura,[65] a Jambi-based business man born in the village of Seponjen, received a forest conversion permit from the district government of Batang Hari (*Izin dari Kabupaten Batang Hari Nomor: 593.3/5567/pem.tgl 22 desember 1992*). In the following years, a number of organized forest conversion, drainage and settlement projects were conducted in the lower and flood-prone peat swamps, some of them overlapped with the borders of the forest reserve TAHURA Sekitar Tanjung. The largest forest conversion initiative started in 1997 and provided land for Bugi migrants. Today, the settlement is one of three hamlets of Seponjen. The hamlet is colloquially called Dusun Bugis. In Kampung Laut village expansion started in the 2000s.

The 1990s and the early 2000s saw an unprecedented logging boom in the Batang Hari delta and in the forests that later became part of the Berbak Carbon Initiative. After the fall of Suharto, in the late 1990s and in the early 2000s, villagers in Kampung Laut and Seponjen were able to receive IPKRs (small-scale logging permits) issued by the *Bupati* of the district of Muaro Jambi.[66] Villagers in Seponjen received small-scale logging permits and formed logging groups that consisted of up to ten individuals.[67] Logging occurred outside and within the state forest. In both villages, logging was a very significant income source, which attracted migrants and led to the establishment of sawmills. In 2005, three years after the enactment of Regulation 34/2002 that prohibited logging permits issued by district authorities, large anti-logging raids by the forest police stopped most commercial logging activities in the villages.[68]

Since the *Hutan Suaka Alam* Berbak became a Ramsar Site in 1991 and a national park in 1997, the region has been a target area for a number of conservation initiatives aiming to develop an integrated spatial planning concept for the different land-use categories and jurisdictions of the region (Lubis and Suryadiputra, 2004: 111–115). The first larger initiative was the Integrated Swamps Development Project (ISDP) running from 1997 to 2000. ISDP was funded by the Global Environmental Facility (GEF) and implemented, among others, by Wetland International, the Provincial Government of Jambi and the MoF. The main output of the project was a management plan for the Berbak National Park and its surroundings. The plan recommended to not extend the logging concessions south of the national park and to designate areas for community forest concessions (ibid. 113). The follow-up project, the GEF Ber-bak-Sembilang Project (2000–2004), was again an attempt to develop a more integrated spatial plan (ibid. 114). In parallel, the first initiative explicitly linking conservation and especially peat-land conservation with climate change miti-gation started in the early 2000s (ibid. 115). The three projects had components that involved local communities. In Kampung Laut, key informants mentioned that they received support for planting trees on idle village land outside of the state forest.[69]

In any case, the various attempts to establish a larger protected area as a new scale of regulation to protect the different ecosystems have been challenged by many actors. The province of Jambi is developing an international seaport and industrial zone in the Batang Hari delta region, only a few kilometers north of Berbak National Park. The MoF has ignored the recommendations of the ISDP project, extended the logging concessions, and failed to ensure that communi-ties receive community forest concessions. State agencies such as the Transmi-gration Agency of the District of Muaro Jambi have even supported further peat swamp conversion by developing transmigration settlements within the forest reserve TAHURA Sekitar Tanjung, challenging the authority of the Pro-vincial Forest Agency and of the MoF (see Chapter 5).

Conflicts between local communities and companies running plantation estates or between local communities and central state authorities were not as common and violent in the villages around the Berbak Carbon Initiative as in the villages around the Harapan Rainforest. A potential explanation is that most of the land claimed as village land in Air Hitam Laut, Seponjen and Kampung Laut does not overlap with the state forest territory, implying that the district government has authority over the land. Generally speaking, the district gov-ernments in the region seemed to accept traditional village lands. In addition, the MoF reclassified large forest tracts to non-forest land in the late 1980s in the surroundings of the villages. This has reduced the potential for conflicts. Vil-lage governments had been directly involved in negotiations with companies and were in some cases even actively searching for investors.[70] Only recently, conflicts occurred but they were not associated with displacements or dispos-session; they were rather caused by controversies about the interpretation of benefit sharing and contract farming agreements between oil palm plantation companies and local communities.

Historical and customary rights regime within and around today's Berbak Carbon Initiative

In pre-colonial and colonial times, agriculture was of minor importance in the Berbak landscape (Guillaud, 1994: 177; Sevin and Benoît, 1993: 96). Fishing and extracting timber and non-timber forest products were much more important activities. Key informants in Seponjen claimed that the village has a commu-nity forest and fruit tree gardens, the exploitation of which dated back to their ancestors.[71] Community members did not need a permit from a customary leader to convert the community forest. Land was abundant and since agriculture was not important, access did not need to be regulated for community members.[72]

Things changed in late 1960s. The *Pasirah* of the Marga Berbak facilitated the creation of the Air Hitam Laut village by permitting Buginese migrants to drain and convert peat-swamp forest around the mouth of the Air Hitam Laut River. Until the end of the 1990s, Malay families were allowed to convert forest according to their needs, with decisions about forest conversion made at the household scale.[73] Starting in the late 1990s, the government of Seponjen village for example restricted individual forest conversions. Decision making about converting the community forest was rescaled, producing a village-scale of regulation. In the following years, the communities undertook more organized, cooperative initiatives to drain and convert peat swamps. In Kampung Laut and Seponjen, the village governments required that farming groups be formed to convert the communities' traditionally owned peat-swamp forest.[74] The formation of farming groups and regulations restricting individual forest conversion permitted Malay elites to consolidate their control of village land and timber resources. Moreover, the draining and conversion efforts were only possible with the collective labor of a farming group.

Malay village communities in Seponjen and Kampung Laut, and Bugis in Air Hitam Laut, have managed to keep control of most of their village territory. The village communities of Kampung Laut and Air Hitam Laut still possess larger tracts of fallow land and secondary forests. They have allocated parts of their land to companies, thereby further strengthening the village-scale of land tenure regulation. However, as in the landscape around the Harapan Rainforest, the ethnic structure has changed: Javanese, Sundanese and Bugis now outnumber the Malay population in Seponjen and Kampung Laut.

De facto land tenure and the "making" of new property in the state forest territory

Agricultural expansion and highly complex local land tenure regimes are an important challenge to conservation and REDD+ projects in Jambi. The rescaling of forest and land governance in the context of the decentralization processes of the *Reformasi* era provided additional agency and the opportunity for local actors to expand their authority over forest and land tenure. Consequently, de facto access and property relations in and around the Harapan Rainforest and the Berbak Carbon Initiative differ substantially from de jure land and forest rights. Large parts of Jambi show similarities with what Peluso (1992) has described as "forests without trees". Peasants and indigenous communities use large parts of state forest territory, including land located within REDD+ and conservation projects, and have transformed forests to swiddens, rubber and oil palm plantations without having de jure rights.

Local public authorities have established a village scale of regulation to legitimize land claims based on an informed choice of regulatory elements of

Indonesian property law, and have started to issue village-level land titles. Actor coalitions involving customary leaders, village governments and district agents have reinterpreted, copied and transplanted the transmigration program and elements of Indonesian property relations to the sub-national scale in order to legitimize forest conversion, land allocation and land transactions within the state forest, including areas that became REDD+ projects after 2008. The various examples outlined in the following show how rescaling has broadened the opportunities for local actors to access land and property. They show that different state apparatuses might act in contradictory ways and not directly reflect the will of dominant groups at the national scale. By using language and symbols of the central state, such as *sporadik* and *transwakarsa*, and by referring to elements of the central state's legal system, actors gain power and legitimacy (Lund, 2006: 687). Local public authorities, e.g. village governments, follow the policies and regulations that specifically support their interests. The different informal settlement projects in the state forest indicate that specific rules or modes of regulation (Etzold et al., 2009: 8) are negotiated between the political elites (village leaders and customary authorities).

The examples illustrate how the power constellations have changed: local public authorities have been able to develop invisible, hidden and visible power. The active reformulation of the "development" and "progress" narratives of the Suharto area has provided new meanings and legitimacy for agricultural expansion and settlement formation, indicating invisible power. Successful occupations of the state forest, like those by the SPI peasant movement, show that the MoF and central state have lost power – thus providing opportunities for alternative developments. In this sense, village-scale land titling and the various village-scale settlements that violate forest law can be seen as active scalar resistance against the central state's land allocation policies, especially those of the MoF.

Land-allocation policies (e.g. towards conservation, oil palm companies and REDD+ project developers) and forest law are considered unjust – particularly in Bungku – and are actively challenged by local actors, as this former village head states: "My house is located within the state land. The state officials say that is state forest, my house is state forest. That means that myself, as a village leader, I am living within the state forest. There is no non-forest land (APL) available [. . .]. We cannot get a land certificate for the state forest. We live in this country but we cannot get land rights [. . .]. According to the Forest Law 1999/41, we should be punished. This is unfair [. . .] Many people have been living within the Asialog concession since the 1960s. Now the area [Harapan Rainforest] is designated for being the lungs of the world and for forest restoration [. . .]. There is no place for the rights of the people there. The communities are threatened by the application of the forest law. There is no social justice, no partnership [. . .]". [75]

Illustrating de facto access and property relations: migrants and access to land

Rural migration makes access and property relations visible. Rural migrants are a very heterogeneous group and they have used many different strategies to access land. Some just migrated a few miles from a neighboring village while others came from areas as far away as South Sulawesi in Indonesia's "Far East". The term *migrant* is a rather blurred category in a dynamic rural frontier landscape. Especially in the landscapes of today's Harapan Rainforest, where semi-nomadic groups dominated until the 1970s, many families are migrants, to some extent. Most migrants who arrived were landless and managed to gain access to land with the support of family members and by engaging with the indigenous population and local public authorities. Furthermore, the common and ongoing process of rural migration in dynamic, tropical frontier landscapes is a major challenge for conservation initiatives seeking to conserve a landscape's status quo (Carr, 2009: 356; Hein and Faust, 2010: 12; Zelli et al., 2014: 29).

The remaining forest within the Harapan Rainforest and the peat swamps of coastal Jambi around the Berbak Carbon Initiative are important destinations for frontier migrants. As in other frontier regions (Li, 2005: 136; 2002: 429), migrants had to engage with the indigenous population and local public authorities in order to convert forest and access land (Figure 3.3).

A Javanese migrant who came to Bungku in the 1970s explained: "I directly started to work together with the Orang Kubu[76] in the forest. We were looking for *jernang*,[77] rattan, timber and other forest products. I worked for 12 years with them in the forest and in 1975 I married my woman who is from here and we stayed in the village [. . .]. After, the forest products were getting scarcer [. . .] I started farming in 1995".[78] Local customary regulations require all migrants to obtain permits from a local authority (a village head or customary leader), or at least from a member of the community. A former community leader in Bungku said, "Newcomers should first contact the village head or the *ketua adat* to ask for permission to clear forests".[79] A former community leader in Tanjung Lebar confirmed: "Forest conversion had to be reported to the village head or at least to an elder [. . .]. Migrants cannot just convert forest; the forests had been already used by our ancestors".[80] A migrant from Kerinci (located in the Bukit Barisan mountain range of western Jambi) said that he asked the Batin Sembilan before establishing his plots inside the borders of today's Harapan Rainforest project.[81] Various local authorities complained that recently many migrants and outsiders had been converting forests without permits. A customary leader from Tanjung Lebar explained: "Those who convert two to three hectares still

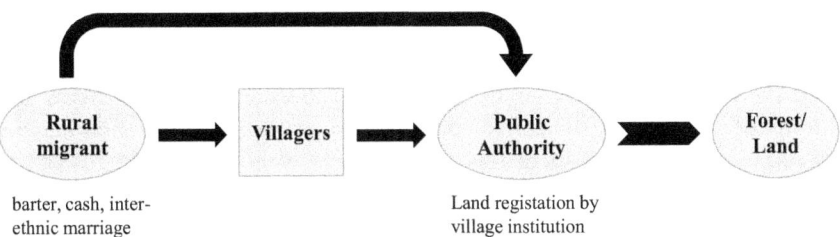

Figure 3.3 Migrants accessing land in the villages

(Source: Hein, 2013b: 17)

have a permit from us – from the Batin Sembilan – but those that have financial resources convert 20 to 50 hectares of land without asking us".[82]

To access land in the villages around Harapan Rainforest, migrants usually had to compensate the land's previous user for his losses instead of buying the land outright, a transaction which was not considered to be a purchase: "The selling of land is actually not allowed, but migrants can obtain land through compensation payments".[83] In the past, cigarettes, rice and other in-kind benefits were sufficient to access land and to get a permit to clear forests from a Batin Sembilan elder or village head.[84] In most cases today, cash payments have to be made that are substantially larger than the earlier IDR 200,000 per ha – and can go up to IDR 70 million[85] per ha for land planted with harvestable oil palms.[86] In recent years, migrants and locals have started to abandon the concept of compensation, and now describe the transactions as "purchases". A migrant from Kerinci who was farming inside a forest reserve stated: "I bought the land and converted the land and planted oil palms [. . .] I have paid IDR 4 million per hectare".[87] A Batin Sembilan woman said that she had sold land: "We sold 3 hectares to family members living in Jambi city because we had to cover the costs of a medical treatment".[88] Some migrants reported that they did not have to pay for the land but they did have to pay for the public infrastructure in the settlements, as well as for land measurement services and land titling.

Among the villages around the Berbak Carbon Initiative, it was only in Kampung Laut that the concept of compensation was known. According to other informants, land trade was even supported by the village government of Kampung Laut. In line with narratives used by the central state in Jakarta to legitimize the transmigration program, the village government officials argued that migrants promote village development.[89] There, land commodification seems generally more advanced than in the Harapan Rainforest area. Key informants in Kampung Laut and Seponjen confirmed that many native Malay have sold their land to migrants, companies, or outsiders.[90]

In both regions, selling land to migrants accelerated processes that transformed large parts of the local indigenous population from property owners who controlled vast forest areas to marginalized smallholders or even to landless peasants. Tania Murray Li (2002) discovered similar dynamics in Central Sulawesi. Within the Harapan Rainforest, the Batin Sembilan facilitated forest conversion and land access for migrants, but as a PT REKI survey (2011a) shows, only a minority of the Batin Sembilan has been able to maintain their land and only the elites have benefited from land trade.

The reasons for selling land are hard to identify. One explanation might be that the owners lack economic resilience because they lack other assets and knowledge about how to act in a commodified, market-oriented environment. In some cases, land was sold to overcome shocks – the costs of medical treatment or the children's education, harvest failures or the destruction of a plantation by forest fires. Another relevant point might be that land had always been abundant and the Malay and Batin Sembilan groups were not yet able to adapt to the new scarcity.

Village-scale land-titling and the role of local public authorities

Evidence for the territorial claims of local public authorities in Jambi is provided by the fact that village governments issue and allocate village-scale land
titles in particular to migrants. Although village-scale land titles have no clear
legal basis, they are common all over Jambi and Indonesia (Kunz et al., 2016).
Village-scale land titles are based on a place-specific interpretation of elements and fragments of state laws and regulations (ibid.). As explained, Governmental Regulation 24/1997 is relevant in this context because it explicitly
mentions the role of village governments in the land-titling process. It gives
village heads the authority to issue specific documents required to transform
customary land claims into a certificate of the National Land Agency (*Badan
Pertanahan Nasional*). The regulation might thus provide legitimacy for local
public authorities (e.g. village governments) to consolidate control over untitled customary land. Moreover, the regulation provides legitimacy to facilitate
land transactions and room for political and economic rent-seeking for village
governments (c.f. Lund, 2008: 23).

Conceptually, village-scale land titles might be seen as the result of rescaling
processes. The day-to-day interactions of smallholders and rural migrants seeking to get land certified by village governments, and local banks' acceptance of
some village-scale land titles as collateral for small loans, consolidates the village
scale of land tenure regulation. Moreover, I argue that the new village scale of
regulation reflects the requirements of a specific mode of production: smallholder oil palm cultivation. Village-scale land titles have facilitated the expansion of smallholder oil palm plantings and contributed to the commodification
of former lineage-based property.

Currently, at least seven different documents issued by the village authorities
or by sub-district authorities are used in the different villages surrounding the
Harapan Rainforest and Berbak Carbon Initiative (Table 3.4). The most common important one, *sporadik* (Figure 3.4), will be explained in detail.

According to Governmental Regulation 24/1997, *sporadik* is the National
Land Agency's application procedure for an individual land title. However, in
the villages I studied and in other parts of Jambi, *sporadik* refers to a village-scale
land title and at the same time to a document that can be used for the National
Land Agency's official *sporadik* titling procedure. *Sporadik* has different meanings
at different scales: at the village scale, the document is a land title that provides
security of tenure; at the national scale, the National Land Agency considers it
to be one of many documents required when applying for a land title through
the *sporadik* titling procedure (Figure 3.4).

The *sporadik* village-scale certification process is facilitated by the neighborhood head (*Ketua Rukun tetangga (RT)*), who organizes the certification
process and appoints a team that usually consists of landowners of adjacent
plots and village elders; in some cases, the head of the hamlet, village secretary
and customary leaders are also involved. The team approves the plot size and
borders, and verifies that no other party claims it. Once the team has approved

Table 3.4 Village-scale land titles used in the study villages

Land title	Used in:	Scope	Authorities involved	Issued within REDD+ project	Scale of regulation
Surat Keterangan Tanaman Tumbuh (SKTT)	Bungku	Certifies cultivation rights/rights to plantings	Village head and village secretary	Yes, Harapan Rainforest	Village
Surat Warisi	Kampung Laut	Certifies land inheritance	Village head	No information	Village
Surat Jual Beli	Bungku, Seponjen	Certifies land transactions	Neighborhood head (*Ketua RT*)	No information	Neighborhood (RT)
Surat Tanah Hak Milik	Bungku	Certifies land ownership	Village head	Yes, Harapan Rainforest	Village
Segel	Kampung Laut	Certifies land ownership; precursor of *sporadik*	Village head	No information	Village
Sporadik Desa	Air Hitam Laut, Bungku, Seponjen, Kampung Laut, Tanjung Lebar	Certifies land ownership; entered into village land registry	Village head, neighborhood head (*Ketua RT*), and/or head of drainage canal (*Ketua Parit*)	Yes, Harapan Rainforest and Berbak Carbon Initiative	Village
Sporadik Kecamtan	Seponjen	Certifies land ownership; entered into village land registry	Sub-district head (*camat*), villhead village head, neighborhood head (*Ketua RT*), head of drainage canal (*Ketua Parit*)	No	Sub-district

(Source: compiled by the author)

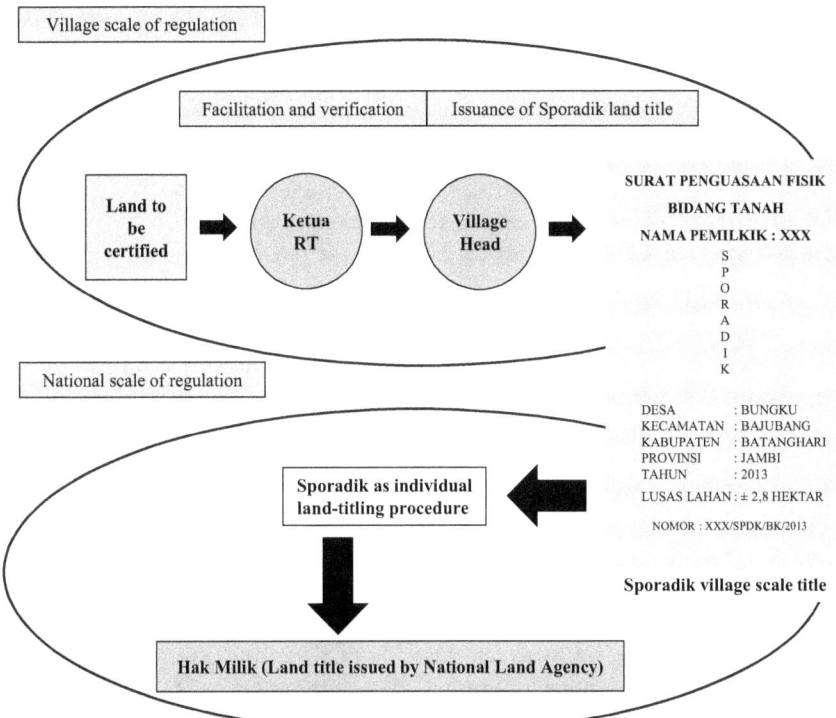

Figure 3.4 Village-scale land-titling in the villages and the changing meaning of *sporadik* across scales

(Source: compiled by the author)

the ownership rights, the village head issues the *sporadik* title. This certification procedure is quite similar in the various villages. The Bungku village government, for instance, issues *sporadik* land titles also for plots in state forests and even for plots in the Harapan Rainforest (c.f. Mardiana, 2014: 42). In Seponjen, two types of *sporadik* titles are in use: titles issued by the village government and those jointly issued by the head of the sub-district (*camat*) of Kumpeh and the village government. Both types are considered to be equally strong, and both can be used as collateral for bank loans and the formal land-title application process.[91] The titles issued jointly by the sub-district head and the village government might reflect Bugi migrants' higher expectations of formalized land titling. As newcomers, Bugis do not belong to local kin networks that provide access to land and tenure security (c.f. von Benda-Beckmann and von Benda-Beckmann, 1999: 18). They are, therefore, interested in getting strong land titles. The document might also be traced back to the role of the *pasirah* in the *margas* of Berbak and Kumpeh, who allocated drainage and forest conversion permits to Bugi migrants. The position of a *Pasirah* and the administrative unit of a *marga* were abolished through administrative reforms and replaced by the *kecamatan* (sub-district), in which the *camat* replaced the *Pasirah*. At least officially, the *camat* has no authority over land. However, as the sub-district's highest public authority who supervises village heads and the *Pasirah*'s successor, his signature might help to increase the legitimacy of *sporadik* titles.

Counter territories and settlement schemes prior to the formation of the Harapan Rainforest project

As mentioned, during the period of political turmoil local public authorities, mainly village and district governments, started to reclaim the state forest territory and organized forest conversion, settlement formation and village expansion. Before the start of the Harapan Rainforest project in 2010, at least four settlements were established within today's project area (Figure 3.1C, D and Table 3.5). The settlements' formation was facilitated by actor coalitions (Figure 3.5) involving Batin Sembilan elites, members of village governments, and migrants – most of whom were Javanese. The coalitions received support from local investors, members of district and sub-district governments, and peasant organizations such as SPI and STN (Serikat Tani Nasional) (c.f. Hein et al., 2016; Mardiana, 2014; Silalahi and Erwin, 2013). The formation of the settlements of Transwakarsa Mandiri (TSM) (Bungku), Camp Gunung (Bungku), Tanjung Mandiri (Tanjung Lebar), and Sungai Jerad/SPI settlements (Tanjung Lebar) led to the conversion of approximately 14,000 ha of forest in the Harapan Rainforest by 2013 (REKI, 2013). Especially in Bungku, settlement formation has been further legitimized through the issuance of village-scale land titles such as *sporadik* and SKTT.

The actor coalitions actively reproduced the New Order's development narratives to justify forest conversion and settlement formation in the state forest. Key informants emphasized that the settlements were intended to provide land for landless migrants and agricultural extension services for poor Batin Sembilan families to help them to overcome "backwardness" and achieve "development".[92] The settlement projects mostly provided peasants with land. Regulations developed by Batin Sembilan elites, village governments, and Javanese migrants limit the maximum amount of land per household and stipulate that direct replanting must occur after forest conversion.[93]

The regulations incorporate elements of formal state law, such as the Basic Agrarian Law (maximum amount of land) and the Forest Law (prohibition of oil palm cultivation in state forests), and customary regulations (replanting after forest conversion). The institutionalization of the settlements (establishing complex rules) and legal mimicry (c.f. Kunz et al., 2016) (using state language and elements of formal state law) further legitimized the settlements. Moreover, the settlements indicate different forms of scale jumping, for instance, village and Batin Sembilan elites jumped down to the district or sub-district scale to obtain forest conversion permits and circumvent the MoF. For their part, migrants

Table 3.5 Settlements within the Harapan Rainforest

	Transuwakarsa Mandiri (TSM)	Camp Gunung	Tanjung Mandiri	SPI Settlements: Sungai Jerad and Bukit Sinyal
Area (ha)	1731	2073	6334	2500
Population*	111 households	302 households	At least 1500 households	At least 508 inhabitants (only Bukit Sinyal, no data for Sungai Jerad)
Hamlet	Kunangan Jaya I	Kunangan Jaya II	Tanjung Mandiri	Pangkalan Ranjau/Mang-kubangan
Village	Bungku	Bungku	Tanjung Lebar	Tanjung Lebar
Sub-district	Bajubang	Bajubang	Bahar Selatan	Bahar Selatan
District	Batang Hari	Batang Hari	Muaro Jambi	Muaro Jambi
Initiated in	2003–2004	2002–2004	2003–2006	2007–2009
Local authorities	Village head, customary leaders	Village head, customary leaders	Village head, customary leaders, *camat*	Hamlet head and SPI leaders
District authorities	Agricultural Agency, Education Agency, *Bupati***	Agricultural Agency, Education Agency, *Bupati***	Agricultural Agency, Education Agency, *Bupati*	No
Settlers hold village-scale land titles	Yes	Yes	No	No
Regulation on max. agricultural land per household	5 ha	No	3 ha	Yes: In 2007 4 ha; now 6–10 ha, depending on household size
Rules for cultivation practices	Direct replanting after conversion	Direct replanting after conversion	Direct replanting after conversion	Oil palm cultivation prohibited, direct replanting after conversion

(Sources: Dinas Kehutanan Kabupaten Batang Hari, 2012; REKI, 2011a, and own investigations, 2013)

* The demographic data is questionable, the official data on village population often indicates a much lower population than village-scale data suggests.

** The actors claim to have received a permit from the *Bupati*.

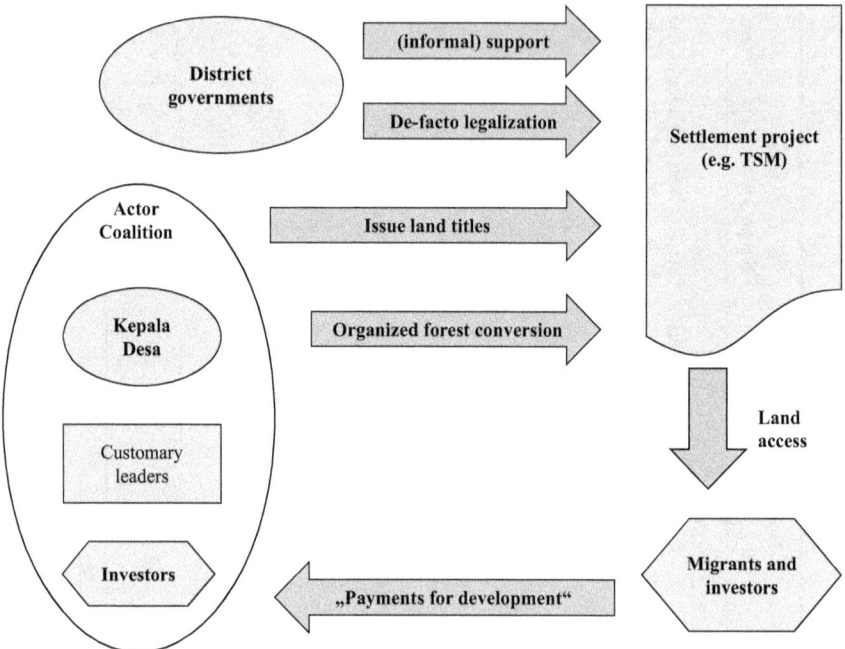

Figure 3.5 Schematic representation of settlement formation in the former PT Asialog concession (now the Harapan Rainforest)

(Source: compiled by the author)

jumped down to the village scale of regulation in order to access land and land
titles, thus stabilizing the village scale of regulation.

Transwakarsa Mandiri / Kunangan Jaya I

As outlined in the conceptual framework of this book, national policies structure *in situ* access and property relations and provide them with legitimacy. The
formation of the Transwakarsa Mandiri settlement in the hamlet of Kunangan
Jaya I of Bungku village can be considered as mimicry of the national transmigration program (c.f. Kunz et al., 2016). A mimicry of policies, legal procedures
and narratives was used to legitimize and justify settlement formation and forest
conversion (Hein et al., 2016; Kunz et al., 2016). The settlement name, Transwakarsa Mandiri (TSM), directly refers to the earlier central-state *Swakarsa*
program, a sub-program of the transmigration program. The New Order era
Swakarsa program provided subsidies to relocate landless migrants from Java,
Madura and Bali, as well as land titles from the National Land Agency (Fearnside, 1997: 555). However, the TSM settlement in Bungku has no official connection to the *Swakarsa* program.

The TSM settlement was founded in 2004 by a Batin Sembilan leader living in Muaro Bulian, the district capital, a Javanese teacher named Pak Kumis[94] and a former Bungku village head. The latter had married into a Batin Sembilan family and claims to represent formal village and customary authority. First, the three leaders requested a permit from the PT Asialog logging company, which was refused, as the company argued that such matters were the MoF's purview. However, the three leaders never received a formal permit from the MoF, or from the forest agency of the district of Batang Hari, or from a transmigration authority.[95] Nevertheless, Pak Kumis claimed: "[. . .] We received a permit from the district head to plant rubber".[96] While it is impossible to verify his claim, it does appear likely that district officials supported the forest conversion and settlement formation since the Agricultural Agency and District Education Agency later de facto legalized the settlement. The Agricultural Agency provided agricultural extension services, such as allocating fertilizer, soy and corn seeds for the settlers. Today, the operational support that the settlement's elementary school receives from the District Education Agency further legitimizes the settlement.[97] The village government of Bungku issued village-scale land titles (*sporadik* and SKTT) to legalize individual land claims.

According to the settlers I met, the TSM settlement had three objectives which were perfectly in line with the objectives of the official transmigration program and with those of the program for underdeveloped villages. The settlement provided land, welfare and employment for poor and landless peasants and aimed to support poor Batin Sembilan households. Pak Kumis said, "In 2004, we requested land for a farming group to support 52 Batin Sembilan households, building houses for them, educating them".[98] The programs also aimed to sedentarize the Batin Sembilan and teach them "modern farming techniques" which would help them to achieve "development".[99] As in the formal transmigration program, migrants participating in the TSM program were supposed to act like model farmers, thereby convincing the "backward" Batin Sembilan to abandon shifting cultivation. However, few Batin Sembilan families were able to participate in the program.[100]

To access land, migrants had to pay a development or administrative fee of approximately IDR 700,000–1,000,000 per ha (equivalent to US$ 55 to US$ 80). Households were allowed to own a maximum of 5 ha of cropland. The development or administrative fee was intended to finance the settlement's infrastructure, including roads, electricity supply, housing for Batin Sembilan, and an elementary school. The organizers of the TSM settlement used the term "development fee" to obscure the fact that land was actually being sold. Participating settlers reported how Pak Kumis promised them that, as in many transmigration settlements, a plantation company would develop a smallholder scheme[101] – but no plantation company ever started to operate there.

In 2007, the District Forest Agency and the forest police intervened. Pak Kumis was arrested for illegal logging and convicted by the Muaro Bulian district court to a prison sentence of one year. In 2010, the PT REKI conservation company received a conservation concession that overlapped the TSM settlement. The community's request to exclude the settlement from the state forest and from the conservation concession was refused by the Provincial Forest Service.[102] A detailed description of the subsequent land conflict involving the settlers, forest service, and the Harapan Rainforest project is given in Chapter 5.

Camp Gunung/Kunangan Jaya II: land for second-generation transmigrants

Camp Gunung is the second forest conversion and organized settlement project initiated by the government of Bungku village and Batin Sembilan elites that challenges the integrity of the Harapan Rainforest project. Like the TSM settlement, Camp Gunung predated the Harapan Rainforest project. PT AAS (PT Agronusa Alam Sejahtera) and PT REKI received their concessions in 2009 and 2010, respectively, and with the support of the police began to campaign against the settlers – in a way reminiscent of the TSM case (more details in Chapter 5).

Camp Gunung is part of the hamlet of Kunangan Jaya II (part of Bungku village), and is located in a border triangle in the state forest between PT REKI's conservation concession (Harapan Rainforest) and the timber plantation concessions of PT Wanakasita Nusantara and PT AAS. Formed between 2002 and 2004, Camp Gunung can be traced back to the deceased Batin Sembilan leader, Pak Yamin Almarhum and to the former village head of Bungku.[103] The village government permitted forest conversion within the former PT Asialog concession and supported the settlement's formation financially.

A key informant explained that Pak Yamin Almarhum had a land conversion permit from the district government of Batang Hari.[104] As in the TSM case, it is impossible to verify this claim, but in the early 2000s, district governments did indeed commonly issue forest conversion permits (c.f. Barr et al., 2006: 2). Later, the settlement was legalized de facto through agricultural extension services provided by the Agricultural Agency of the District of Batang Hari, and also by the establishment of a school and a kindergarten financed by the Education Agency of the District of Batang Hari.[105]

From the late 1990s to the early 2000s, Bungku became an important destination for second-generation, mostly Javanese, transmigrants from the Batang Hari Delta transmigration settlements of Nipah Panjang and Ranto Rasau in the district of Tanjung Jabung Timur.[106] Pak Yamin Almarhum facilitated land transactions between migrants and Batin Sembilan families, and offered 5 ha parcels of land for IDR 750,000.[107] As in the case of the TSM, the payment was considered to be a development and infrastructure fee, not payment for land.[108] The village government of Bungku and Pak Yamin Almarhum organized the construction of roads and bridges, as well as a school, mosque and rice mill.[109]

Tanjung Mandiri: migrants as model farmers and conflictive boundaries

Tanjung Mandiri is the largest settlement in the Harapan Rainforest. Forest conversion and settlement formation started there in 2003; activities intensified in 2006 with the construction of houses and the establishment of the first plantations.[110] Today, the settlement is one of five official hamlets of Tanjung Lebar. Like Camp Gunung and the TSM settlement, Tanjung Mandiri grew out of agreements involving customary elites and local branches of the government, in this case the village government of Tanjung Lebar and the sub-district head (*camat*) of Sungai Bahar.

As in Bungku, the former village head of Tanjung Lebar, a Javanese migrant who had married into a Batin Sembilan family, claimed to represent both customary and village authority. The Tanjung Mandiri settlement, and especially the way in which the Tanjung Lebar village elites framed and legitimized it, has a lot in common with the TSM settlement of Bungku. Key informants in Tanjung Mandiri framed the settlement project as a "win–win" situation for migrants and the Batin Sembilan. Village elites in Tanjung Lebar also reproduced the narratives used by the New Order regime to legitimize the transmigration program and argued that Batin Sembilan would benefit from the commercial farming techniques introduced by migrants.[111] A key informant explained that Tanjung Mandiri "[…] is based on an agreement between SAD[112] and migrants with the aim of developing the SAD that have not understood how to farm yet. The settlement project provided the opportunity to meet migrants with modern farming techniques so they can work together on the land that was already used by their ancestors".[113] Village and Batin Sembilan elites described Tanjung Mandiri as a settlement project jointly organized by Batin Sembilan and migrants. A settler explained: "[…] The settlement was organized by Pak Adin[114] acting on behalf of the SAD. Without involvement of the SAD, we would not have dared to settle here because it is part of their land. He [Pak Adin] organized from the beginning that people settle here".[115]

Land settlers had to pay a "measurement fee" to access land. As in Bungku, key informants stressed that the land as such was gratis. The measurement fee and settlers' additional voluntary contributions were used for infrastructure development, especially to construct a school, village hall, mosque and church. The first settlers received 2 ha of cropland and 1 ha for a house, yard and garden. Settlers were not allowed to own more than 3 ha of land, and land had to be planted immediately after forest conversion.[116]

Today, the settlement has been de facto legalized by the Education Agency of the District of Muaro Jambi, which supports its school.[117] In 2011, the head of the Muaro Jambi district strengthened the settlers' position by celebrating the traditional rice harvest festival in Tanjung Mandiri and promising that the settlement would no longer be part of the Harapan Rainforest.[118] However, a district head has no formal authority to reclassify state forest, and the area is officially not even part of the district of Muaro Jambi. According to MoF maps, the Harapan Rainforest concession is located in the village of Bungku in the district of Batang Hari and not in the district of Muaro Jambi. Nevertheless, the northeast of the Harapan Rainforest concession is part of customary land (*wilayah adat*) belonging to Batin Sembilan of Tanjung Lebar, which has been part of Muaro Jambi district since 1999. Consequently, most settlers identify as being from Tanjung Lebar – not from Bungku.[119] Batin Sembilan from Bungku and the village government of Bungku question the land claims of the Batin Sembilan of Tanjung Lebar. Batin Sembilan in Bungku accuse Batin Sembilan elites in Tanjung Lebar of selling land that traditionally belongs to groups from around Bungku.[120]

Agrarian reform through land occupation

The SPI settlement project is located south of the pre-colonial hamlet of Mangkubangan/Pangkalan Ranjau and can be considered as being part of the organization's multi-scalar campaign for agrarian reform (see Chapter 5). The settlement is the most recently founded settlement, and the most contested settlement in the Harapan Rainforest concession (Hein and Faust, 2014; Hein et al., 2016). An SPI activist claimed that SPI members have been present in the area since the late 1990s.[121] One of the first SPI members to live in the area was a teacher who had married into a Batin Sembilan family and who swapped a motorcycle for land from the *Temenggung* Seman who controlled the area around Sungai Jerad.[122] *Temenggung* Seman and the teacher were also involved in initiating the SPI settlements. The formation of the SPI settlements started in 2007 after SPI's flag was symbolically hoisted in the presence of the hamlet head of Mangkubangan/Pangkalan Ranjau. However, some members of the village government complained that SPI started the settlement project without having formally asked the village head.[123]

As in the other village-scale settlement projects, reference was made to the transmigration program. The head of an SPI basis argued: "During the Suharto era we had the transmigration program paid by the government: trillions had been paid by the government for transmigration. Today it's different [. . .]. Today farmers like us, we have the problem that we do not have an official permit from the government".[124] In 2013, the area claimed by the SPI in encompasses more than 2500 ha of PT REKI's concession and has as many as 18 smaller settlements, each inhabited by up to 40 households. Recent numbers claim that SPI occupies 17,000 ha (Jambipos Online, 2017).

The settlements have a complex institutional structure established by SPI, which has divided its territory into blocks of land of 50 to 60 ha, with each block used by approximately ten households. Not all blocks had been converted during field research in 2013. Four to five blocks form a "basis", the lowest level of political organization in the SPI. Each basis is led by an elected head (*ketua basis*) and has 2 ha of collective land that is planted with rubber. To access land, smallholders must have a residence permit issued by the village government of Tanjung Lebar, belong to SPI, be landless, and pay an IDR 300,000 (equivalent to US$ 25) "measurement fee" for each hectare.[125] Land access is facilitated by the SPI leader of the Bahar region (Mardiana, 2014: 36). Part of the measurement fee goes to Batin Sembilan elites (ibid. 51). SPI leaders claim that they only accept poor and non-commercial farmers, which is hard to verify. A local SPI leader stated: "The poor farmers living in this neighborhood (RT) [. . .] came to survive. They have nothing outside, 99 percent of the farmers living here are really poor, they don't have another place of living [. . .] We don't have capitalist farmers living here [. . .] Those are the people I know, maybe there are others".[126]

SPI permits each household a maximum of 6 ha of land; for larger households, up to 10 ha are permitted.[127] SPI leaders stated that a basic rule for all settlements is the three "T" (*tebang, tanam, tumbuh*) rule: forest conversion, planting and growing.[128] Around 2010, after PT REKI got the conservation concession, SPI leaders imposed a ban on oil palm cultivation in the settlements. The oil palm ban indicates that SPI leaders are willing to obey the Forest Law (cultivating oil palms is prohibited in the state forest) and the conservation regulations of PT REKI, and to act in line with the global anti-bio fuel and food sovereignty campaigns of La Via Campesina.[129] According to SPI leaders, settlers who violate the oil palm ban are expelled from the organization; settlers who had already planted oil palms are supposed to replace them with rubber.

Initially, the SPI settlement also aimed to provide benefits for Batin Sembilan families. Mardiana (2014: 51) found that Batin Sembilan elites had not fairly allocated the payments received from the SPI to other Batin Sembilan. Most Batin Sembilan families had not benefited from the land transactions. In July 2013, a Batin Sembilan leader living outside the SPI area said that his groups would like to cooperate with SPI to better understand modern farming techniques.[130] At the same time, he criticized the rapid expansion and forest destruction caused by SPI members.[131]

Village-scale peat-swamp conversion and settlement schemes in the surroundings of the Berbak Carbon Initiative

Village-scale peat-swamp conversion and settlement projects in the surroundings of the Berbak Carbon Initiative have mainly been developed outside of the state forest, that is, outside areas protected in the Berbak Carbon Initiative. However, some initiatives to convert peat-swamp forests extended into the border of the Berbak National Park.[132] In addition, according to conservationists from the ZSL, the conversion of drained peat swamps adjacent to the Berbak Carbon Initiative also affects the hydrological balance of the swamps within the project area and increases the risk of peat fires. In contrast to the settlement projects in the Harapan landscape, the area used for agriculture within the Berbak Carbon Initiative is rather small, with only a few households living permanently within the protected areas. The sole exception is the district-to-district transmigration settlement in the forest reserve Sekitar Tanjung, established by the Transmigration Agency of Muaro Jambi. In 2008, a district-to-district transmigration settlement for Javanese families was established in Kampung Laut – in the Sekitar Tanjung forest reserve and the Berbak Carbon Initiative. The formation of the transmigration settlement in Kampung Laut started in late 2008, with the district government selecting a location that was part of the Forest Reserve Sekitar Tanjung and the Berbak Carbon Initiative (Figure 3.2C, D). This was thus an ineligible site for a transmigration settlement according to Forest Law, and consequentöy the MoF and Provincial Forest Agency attempted to stop the settlement's formation.[133] Chpater 5 contains a detailed description of this land conflict that involved different state apparatuses and settlers.

Most of the peasants I met in Seponjen, Kampung Laut and Air Hitam Laut said they accept the prohibition on planting crops in the state forest territory, and most village governments considered new forest conversion that violated the borders of the Berbak Carbon Initiative to be illegal. A key informant in Kampung Laut said that the village government refused to issue the documents necessary for an identification card for a household that had settled in Berbak National Park.[134] Nevertheless, especially in Kampung Laut, villagers questioned why the forest reserve remains protected: "[. . .] the forest reserve is for protection, but what do they want to protect? There is no timber anymore and animals cannot live there anymore".[135]

It may be that the state forest border is so broadly accepted because after 1985 the release of large areas from the state forest provided enough land for smallholders' agricultural needs. A second explanation could be that converting peat-swamp forest requires large investments that especially Bugi elites were only willing to make if the land's status guaranteed safe returns. A third explanation might be that enhanced law enforcement against illegal logging was initiated by the former president Susilo Bambang Yudhoyono (SBY), preventing the conversion of logged-over forest into agriculture plots. A fourth factor could be that agriculture was traditionally of minor importance for Malay groups in the area; its relevance having only grown very recently. In the past, logging and gathering non-timber forest products were much more important. These activities continue – also within the Berbak Carbon Initiative.[136]

I have identified two different types of organized, collective village-scale forest conversion and settlement initiatives in the villages adjacent to the Berbak Carbon Initiative that I studied (Table 3.6). The first is directly associated with the immigration of Bugis from South Sulawesi, and the second is driven by local Malay groups and Javanese migrants.

Bugi colonization around Berbak National Park

In the 1950s, Bugi immigration grew; in the 1960s, the immigrants started to drain and found settlements in the Berbak landscape (Claridge, 1994: 290; Sevin and Benoît, 1993: 102). The village of Air Hitam Laut dates back to a Buginese drainage and settlement initiative. In the 1970s, the *Pasirah* of the *Marga* of Berbak gave the Bugi the right to drain and convert the peat-swamp forest around the mouth of the Air Hitam River. A key informant in the village explained: "The Bugi leader Pak Janggut[137] obtained the right to drain and convert land from the *pasirah*. Part of the converted land was used to pay the *pasirah* afterwards".[138] Pak Janggut, initiator and head of the project was appointed the "head of the river" (*Kepala Sungai*), and sub-divided land received from the *Pasirah* to his clients called "canal heads" (*Kepala Parit*). Usually the plots were

Table 3.6 Organized peat-swamp-forest conversion initiatives in villages adjacent to Berbak Carbon Initiative

	Village of Air Hitam Laut	Dusun[1] Bugis	Ketapang	Farming groups in Seponjen	Farming groups in Kampung Laut
Hamlet	All hamlets	Sungai Lais	Ketapang	Hamlets I + II	All hamlets
Village	Air Hitam Laut	Seponjen	Kampung Laut	Seponjen	Kampung Laut
Sub-district	Sadu	Kumpeh	Kumpeh	Kumpeh	Kumpeh
District	Tanjung Jabung Timur	Muaro Jambi	Muaro Jambi	Muaro Jambi	Muaro Jambi
Initiated in	1965	1996–1997	2000	1995	late 1990s
Overlapping the Berbak Carbon Initiative and/or state forest	Until 1970s	No	No	Only a few plots	Only a few plots
Local authorities	*Pasirah* and Bugi elites	*Camat*, Bugi elites, Malay elites, village government	Village head, customary leader, head of farming group	Village head, customary leader, head of farming group	Village head, customary leader, head of farming group
Tenure arrangement	*Parit*	*Parit*	*Tanah adat* and *parit*	*Tanah adat*	*Tanah adat*
Land access facilitated by	River head and canal head	River head and canal head	Head of farming group and canal head	Head of farming group	Head of farming group
District authorities	–	Land Agency of the district of Batang Hari	Agricultural Agency	–	–
Settlers hold village-scale land titles	Yes	Yes	Yes	Yes	Yes
Regulation on max. agricultural land per household	No	No	2 ha	1.25 ha per farming group member	2 ha per farming group member

(Source: author's own investigations)

1 English: hamlet.

divided by drainage canals. The canal heads were responsible for facilitating the drainage and conversion of the peat swamp in their plots. They also looked for Bugi migrants interested in land and let them convert the forest areas in their plots. A former *kepala parit* in Air Hitam Laut explained that for each 2 ha converted by the migrants he received 0.25 ha, part of which he paid to the river head and the *Pasirah*.[139] In total, 24 main drainage canals were planned along the Air Hitam Laut coastline, partly overlapping the Berbak *Wildreservaat* (*Hutan Suaka Alam*). Only ten of these have been realized.[140] Drainage activities were stopped by the Governor of Jambi[141] but drainage canals still impact the hydrological balance of the Berbak Carbon Initiative, increase the risk of peat fires, and facilitate access for loggers and for actors gathering non-timber forest products such as jelutung.[142]

Many Bugi settlements and forest conversion projects in the Berbak landscape had land tenure arrangements based on various nested authorities, each equipped with a specific scale of regulation. In Seponjen, the hamlet of Sungai Lais, founded in 1997, is characterized by a similar arrangement. Pak Hadji Pattimura, a Malay business man from Seponjen, held a land conversion permit from the land agency of the district of Batang Hari (*Dinas Pertanahan Kabupaten Batang Hari*). With the Bugi leader Pak Selang and the village government, he organized a peat-swamp conversion and settlement project for second-generation Bugi migrants and local Malay households. The leaders received land for investing in the project.[143] As in the Harapan landscape, elders stated that the settlement would contribute to village development and provide land for poor households.[144] Pak Selang acted as river head (*Kepala Sungai*) and allocated land to 10 canal heads,[145] each of whom was authorized to regulate his own parcel of land. The land was converted by migrants – who again were mainly Bugis. Each migrant received half of the land that had been converted; the other half was divided between the canal and river heads, and Pak H. Pattimura. The land managed by the second and third canal heads was designated for local Malay households. Participating Malay households received all the land they had converted.[146] Many of them later sold it to Bugi or Javanese migrants. The hamlet of Sungai Lais was established approximately two to three kilometers from the borders of today's Berbak Carbon Initiative.

Farming groups for collective forest conversion

In the late 1990s, agriculture became more relevant in Seponjen and Sunga Aur. In addition, immigration of Javanese and Bugis increased. In order to meet rising demand for agricultural land, Malay elites started to convert the forests within the boundaries of their customary village land, called *tanah adat*.

Although this area overlaps the Berbak Carbon Initiative, only a few plots are located in the project area today. However, initially peat-swamp conversion also took place in areas that later became part of the project area. Forest conversion was regulated and controlled by the local Malay elite. Community members who sought land first had to join a farming group. Such groups, each consisting of 20 to 60 farmers, were led by local Malay elites (the customary head, *ketua adat*) and were supported by the village governments – especially by the village heads. Each participant received approximately 2 ha of land. In the early 2000s, the farming groups in Seponjen legitimized their activities in the state forest territory by referring to community small-scale timber extraction permits from the district government.[147] Farming groups reduced the cost of converting land for individuals and permitted the conversion and drainage of larger areas. In Seponjen, and from the early 2000s in Kampung Laut, migrants, too, were allowed to access land by joining farming groups. In the Ketapang hamlet (in Kampung Laut), migrants and locals worked together to drain and convert land. Migrants had to pay to participate in farming groups and received the more flood-prone plots.

Summary and preliminary conclusion

State transformation processes induced rescaling processes that altered the dialectical relationship between structure and agency and consequently changed the ability of actors to access land and property. I argue that four subsequent state transformation processes induced rescaling that has altered Indonesia's forest and land tenure governance, namely: colonization, independence, New Order and *Reformasi*. Each period was characterized by specific land and forest tenure regulations providing different opportunities for actors to access forest and agricultural land. Scaling back, in other words investigating past scalar arrangements, demonstrated that historically contingent structural inequality is a persistent feature of land and forest tenure in Indonesia. Establishing the state forest territory can be considered as an attempt to strengthen territorial control and to promote and privatize the exploitation of forest resources and the expansion of agro-industrial estates, and recently even of protected areas. The *Reformasi* and the *post-reformasi* era provided additional agency for marginalized actors to construct alternative scales of meaning and regulation.

Three, at least partly contradictive, state projects of the post-colonial period influenced the scalar configuration and related access and property relations. First, the Basic Agrarian Law (BAL) as the first attempt to establish a new national scale of natural resource governance promoted (probably not intentionally) the commodifcation of land but, at the same time, imposed restrictions on large landholders and proposed land reform. Today, the BAL is relatively weak in legal terms but is still used – especially by peasant organizations such as SPI and KPA – to underpin political campaigns for a more equal distribution of agrarian resources (Bakker and Moniaga, 2010: 188; Hein and Faust, 2014: 25; Rachman, 2011: 54). Second, the Basic Forest Law has led to a reestablishment of the dualistic governance structure, fostering the dispossession of local and indigenous communities and seeking to construct private-sector-led economic growth as a new national scale of meaning. Third, the Village Law has further undermined traditional *adat*-based scales of meaning and regulation. Moreover, the law has imposed new jurisdictional boundaries and created a standardized and homogenous administrative structure across the archipelago.

Land tenure in the landscapes around the two REDD+ projects is contested and this can be considered as one of the main causes for the conflicts that are described in Chapter 5. An important reason for the contestation of property rights are competing authorities. Overlapping and competing property rights (e.g. property rights legitimized by village governments versus those backed by the MoF) indicate conflicts between different state apparatuses and struggles over the constitution and the legitimacy of public authorities. Overlapping and contradictive land and forest tenure shows that what members of society consider to be legitimate property rights change over time and are subject to historically contingent and ongoing societal struggles over concepts and truth (Sikor and Lund, 2010: 6). National laws and regulations structure access and property relations. However, local public authorities strategically follow those laws that support their own interests, explicitly ignoring others. The construction of village scales of meaning and land tenure regulation can be considered as active scalar resistance strategies.

My investigation indicates that access and property relations are explicitly linked to specific scalar arrangements (Table 3.7). Pre-colonial scales of regulation relevant for access and property relations were based on watersheds controlled by specific lineages. In the Harapan landscape, the *Depati* and the *Temmenggung* were responsible for land tenure issues within specific watersheds, and kin relations regulated access to land and property. Rescaling induced by Dutch colonization changed the scalar structure and led to the establishment of a new authority (*Pasirah*) that was in charge of recognizing land claims for the local population, challenging kinship-based property relations. Indonesian independence and the formation of the Indonesian nation state have diversified access and property relations. Central state apparatuses constructed a national

Table 3.7 De facto access and property relations in the REDD+ project regions

REDD+ project region	Access and property relations
Harapan Rainforest	Customary arrangements (customary authority and adat)
	Property rights backed by village officials
	Property rights backed by private actors (conditional land tenure/Harapan Rainforest)
	National-scale regulation (including transmigration)
Berbak Carbon Initiative	*Parit* system (Bugi colonization)
	Property rights backed by village officials
	Provincial-scale regulation (conditional land tenure/Berbak Carbon Initiative)
	National-scale regulation (including transmigration)

(Source: author's own investigations)

scale of meaning and regulations based on national development and narratives about economic growth in order to allocate exploitation and cultivation rights to corporate actors. Corporate actors actively reproduced the national scale of meaning and regulation by requesting these rights from national-scale authorities, such as the MoF. As a national scale of meaning and regulation, the state forest territory significantly reduced the abilities of peasants and indigenous groups to access land and property.

The socio-spatial configuration changed again after Suharto. Village governments, local elites, and customary leaders constructed new village scales to regulate access to land and property. These new village scales have in turn been reproduced by peasant farmers and especially by rural migrants who request land and village-scale land titles. Village scales of regulation – property rights that are based on unwritten village-scale tenure regulations – most notably overlap with the Harapan Rainforest, challenging the project's integrity and inducing land conflicts. Village governments, local elites, and customary leaders have not only legitimized individual land claims but – supported by district-scale state apparatuses – have also formed larger-scale settlement projects that actively reproduce central state resettlement programs and development discourses (e.g. transmigration). Local public authorities translate and reinterpret central state policies and regulations to underpin the legitimacy of their actions (legal mimicry) and combat historically contingent structural inequality.

To further unpack the various abilities of actors to access land and property, it might be helpful to engage more explicitly with power asymmetries. Political ecologists argue that unequal power relations help to explain uneven access to natural resources, including land (Blaikie, 2012; Bohle and Fünfgeld, 2007; Bryant, 1998; Forsyth, 2008). I argue that the ability of different actors to access land and property is mediated by the three dimensions of power (visible, hidden and invisible) (Gaventa, 2006; Lukes, 2005), as outlined in my conceptual framework. The indigenous Batin Sembilan, rural migrants, elites, village governments, and companies such as PT REKI are differently positioned with regard to the three power dimensions.

In general, peasant farmers – including indigenous groups in the study villages – are poorly positioned with regard to the three dimensions of power. They hold fewer material resources than did corporate actors (e.g. less land), often have no land titles from BPN, and are excluded through institutional procedures and regulations (Forest Law 41/1999). Invisible power, or the social construction of meanings or indirect power (Gaventa, 1982: 15) and the internalization of subordination (ibid. 16), further marginalize peasant farmers and indigenous groups. The social construction of the backward shifting cultivator by the New Order regime also contributed to their marginalization.

After Suharto's fall, village governments and Malay and Batin Sembilan elites were able to significantly expand their authority and establish new village scales to regulate access and property (hidden power). Village governments gained power by using the symbols and language of the state. They control land in alliance with Malay or Batin Sembilan elites and can allocate land, one of most important means of production in rural landscapes (visible power). Batin Sembilan elites in particular were able to significantly change the rules of the game by reintroducing customary regulations and making land claims based on social identity (hidden power). The Batin Sembilan elites' ability to change the rules of the game to their benefit was facilitated by their position in the invisible dimension of power. The rise of global discourses on indigeneity and the emerging indigenous rights movement may have changed the social construction of indigeneity from being associated with "backwardness" to being associated with "rights". It thus provided the opportunity to frame land and forest as indigenous territories or customary land (*wilayah adat*). Most migrants interviewed in both regions did not question the authority or land claims of local and indigenous communities. Social identity has become a relevant source of power in both landscapes, to the benefit of Batin Sembilan and Malay elites.

On the one hand, migrants benefitted from alliances with Batin Sembilan elites because of Batin Sembilan land resources, while on the other hand, Batin Sembilan elites engaged with migrants in response to the social construction of the backwardness of shifting cultivators fostered by the Suharto regime. Interethnic marriages helped the Batin Sembilan to benefit from migrants' social capital. Moreover, the alliances provided significant benefits for both parties, especially a better positionality in regard to the visible and invisible dimensions of power. However, those Batin Sembilan that were not part of the elite of the indigenous group, obviously most of them, have been even further marginalized because they were neither able to claim and allocate land (visible power) nor to benefit from the altered notion of indigeneity.

The changed power balance of post-Suharto Indonesia – caused by shifting relationships between structure and agency – provided the opportunity for open peasant resistance (see Chapter 5). SPI members in the Harapan Rainforest significantly benefitted from their alliances with Batin Sembilan elites. The ability to allocate material resources (land) increased the organization's attractiveness, and probably increased its organizational strength as well as its ability to occupy large parts of the Harapan Rainforest (Hein et al., 2016) (Figure 3.6).

The complex and conflictive access and property relations and tensions and contradictions between de jure and de facto land tenure described in this chapter are the dynamic context for the implementation of REDD+ and other market-based conservation instruments, such as conservation concessions discussed in the next chapter. These new conservation instruments challenge the different village-scale settlements located in the state forest territory and in areas that became REDD+ projects later on. Moreover, REDD+ has triggered the formation of a new transnational scale of forest governance and has strengthened the role of private actors in forest, climate and environmental governance.

Figure 3.6 Picture of area used for shifting cultivation inside the Harapan Rainforest

(Source: taken by the author, 2013)

Notes

1 In other sources, the term *Batin* refers to the leader of semi-nomadic, non-Muslim forest dwellers and not to a specific ethnic group (Andaya, 1993: 14).

2 *Kubu* is a derogatory term for many non-Muslim nomadic and semi-nomadic tribes used by the Malay population and colonial authorities on Sumatra.

3 The *pangeran ratu* was chosen by nobles and was in charge of most government affairs (Locher-Scholten, 2004: 45).

4 The *domein verklaring* is mentioned in article 1 of the *Agrarisch Besluit* of 1870. In the Basic Agrarian Law of 1960, the Dutch terms is translated as follows: General Declaration that lands belong to the state.

5 The "outer islands" refer to all Indonesian islands except Java, Bali and Madura, which are called the "inner islands".

6 *Adat*: According to David Henley and Jamie S. Davidson (2007: 1) the Indonesian term *adat* refers to custom or tradition, including notions of order and consensus. The term *adat* is often used for customary law and the term *adat* rights is often used to legitimize land claims based on customary law, in other words land claims not based on colonial or formal state law (ibid.)

7 Despite the establishment of a Sumatran forest service with specific management territories, state forest management in Sumatra remained unspecific and forests remained largely under the authority of local rulers (Barr, 2006: 19–20; Nurjaya, 2005: 46). After independence Jambi initially became part of a newly formed province of Central

Sumatra consisting of West Sumatra, Riau and Jambi (Dinas Komunikas dan Informatika Provinsi Jambi, 2013). In 1957, Jambi became an autonomous province.

8 Interview with a staff member of the Sajogyo Institute, Bogor, 08.10.2013.

9 Interview with a key informant in Tanjung Lebar, 22.07.2013 and interview with a staff member of Sajogyo Institute, Bogor, 08.10.2013.

10 SK. HPH No. 408/Kpts/Um/9/1971 *tanggal 23 September 1971 tentang Pemberian Hak Pengusahaan Hutan kepada PT. ASIALOG,* Decision of the Ministry of Forestry to allocate a logging concession to the company PT Asialog, own translation) (Menteri Kehutanan, 1995)

11 No detailed information available, according to a key informant in Tanjung Lebar the company received a logging concession in 1974, 27.07.2013.

12 Neighborhood units were not mentioned in the Village Law, they were introduced in 1983 through an additional decree from the Ministry of Home Affairs (Warren 1990: 3).

13 Refers here to the customary land of indigenous communities.

14 Interview with a key informant in Air Hitam Laut, 29.09.2102.

15 Interview with a key informant in Air Hitam Laut, 30.09.2012 and a key informant in Kampung Laut, 30.08.2013. Land allocation by village heads is officially not backed by village law but in practice many village heads are involved in issuing different types of land-use permits.

16 Reformasi refers to the democratization period after President Suharto announced his resignation on 21 May 1998.

17 During 1976–1989, 10 percent of the transmigration budget was covered by the World Bank (Fearnside, 1997: 6).

18 Interview with a key informant in Tanjung Lebar, 25.07.2013.

19 Interviews with a key informant in Kampung Laut, 29.08.2013 and with a staff member of Dinas Kehutanan Provinsi Jambi, 27.08.2013.

20 The reforestation fund (*Dana Reboisasi*) is a volume-based fee charge on harvested timber to support reforestation of logged-over forest (Resosudarmo et al., 2006: 62).

21 Interview with a staff member of the Ministry of Forestry, Jakarta, 23.07.2012.

22 According to Governmental Regulation 56/1960, Article 1 the maximum land size permitted per individual depends on the population density and on the type of land (land for wet rice cultivation and dry land). Twenty ha of dry land are permitted in areas with a low population density, in areas with very high population densities, only 6 ha are permitted.

23 Interview with the Working Group on Forest and Land Tenure, Bogor, 08.10.2013 and with a staff member of Burung Indonesia, Bogor, 11.10.2013.

24 Interview with activists of the NGO PINSE, Sungai Gelam (Jambi) 05.09.2013 and with a staff member of Burung Indonesia, Bogor, 11.10.2013.

25 Interview with a staff member of the NGO Amphal, Jambi, 17.07.2013.

26 Interview with a GIZ advisor, Jakarta, 24.07.2012, with a Greenpeace Indonesia activist, Jakarta, 27.07.2012, and a Forest Watch activist, Bogor, 20.07.2012.

27 Interview with Forest Watch activists, Bogor, 20.07.2012.

28 Interview with an AMAN activist, Jakarta, 27.07.2012.

29 Interview with an AMAN activist, Jakarta, 27.07.2012.

30 Interview with a BPN officer, Jambi, 06.09.2013.

31 Also confirmed in an interview with a BPN officer, Jambi, 06.09.2013.

32 Interview with key informants, Seponjen, 15.09.2013 and 13.09.2013

33 Cited laws and regulations: Makamah Konstitusi Republik Indonesia, 2012; Menteri Kehutanan; 2007; 2008a; 2011; 2014; Ministry of Forestry, 2008; Presiden Republik Indonesia, 1960a; 1960b; 1990; 1997; 1999.

34 Interview with a key informant in Tanjung Lebar, 25.07.2014.

35 Interview with a key informant in Tanjung Lebar, 25.07.2013.

36 Interview with a key informant in Tanjung Lebar, 27.07.2013.

37 The most recent figure is from 2005.

38 The village of Markanding is part of the sub-district of Sungai Bahar, it is located approximately 10 km north-east of Bungku.

39 Interview with a key informant in Tanjung Lebar, 26.07.2013.

40 Interview with a key informant in Bungku, 12.09.2012.

41 According to Hidayat (2012: 20–27) there are at least two other versions of a myth of origin of the Batin Sembilan, the second argues that the Batin Sembilan descend from a leader from Palembang (South Sumatra), and the third consists of elements of the first and the second myths.

42 Interview with a key informant in Bungku, 12.09.2012.

43 The position of the *Temenggung* existed among the Orang Rimba of the Dua Belas landscape (Steinebach, 2013a: 49–50), the extent to which the position of the *Temenggung* was known among the Batin Sembilan in colonial and pre-colonial times is at least debatable, argues Steinebach (forthcoming). Probably, the Batin Sembilan only very recently started to call their leaders *Temenggung*.

44 Interview with a key informant, Bungku, 21.09.2012.

45 Interview with a key informant, Bungku, 12.09.2012.

46 Interview with key informants, Bungku, 12.09.2012 and Tanjung Lebar, 21.07.2013.

47 Interviews with key informants, Tanjung Lebar, 25.07.2013, Tanjung Lebar, 25.07.2013.

48 Interview with a key informant in Tanjung Lebar, 26.07.2013.

49 Interviews with key informants in Bungku, 12.09.2012 and 09.07.2013 and in Tanjung Lebar, 23.07.2013.

50 Interview with a key informant in Tanjung Lebar, 27.07.2013.

51 Interview with a key informant in Tanjung Lebar, 23.07.2013.

52 Interview with key informants in Tanjung Lebar, 26.07.2013 and 27.07.2013 and in Bungku, 08.09.2012.

53 Interview with a key informant in Tanjung Lebar, 27.07.2013.

54 *Koompassia excels,* a specific tree species that is often used by wild bees for beehives.

55 Interview with key informants in Bungku, 12.09.2012 and 21.09.2012.

56 Interview with a key informant in Bungku, 12.09.2012.

57 Interview with a key informant in Bungku, 30.07.2013.

58 Interview with a key informant in Tanjung Lebar, 26.07.2013.

59 Interview with a key informant in Tanjung Lebar, 26.07.2013.

60 Interview with a key informant in Tanjung Lebar, 25.07.2013.

61 Interview with a key informant in Bungku, 24.08.2013.

62 Interview with a key informant in Bungku, 08.09.2012.

63 Bugi migrants have conducted larger-scale land drainage across Sumatra's east coast. According to Hanson and Koesoebione (1979, cited in Claridge, 1994: 292) Bugis converted as much forest as the government led-initiatives in the period from 1969–1973.

64 Interview with a key informant in Air Hitam Laut, 28.09.2012.

65 Fictitious name.

66 Interviews with key informants in Kampung Laut, 28.08.2013 and 01.09.2013.

67 Interview with key informants in Seponjen, 10.09.2013 and 12.09.2013.

68 Interview with a key informant in Seponjen, 12.09.2013.

69 Interview with a key informant in Kampung Laut, 31.08.2013.

70 Interview with key informants in Kampung Laut, 30.08.2013 and Seponjen, 13.09.2013.

71 Interview with a key informant in Seponjen, 11.09.2012.

72 Interview with a key informant in Seponjen, 13.09.2012.

73 Interview with a key informant in Kampung Laut, 30.08.2013.

74 Interviews with key informants in Kampung Laut, 30.08.2013, 31.08.2013 and 01.09.2013 and in Seponjen, 08.09.2013, 09.09.2013 and 13.09.2013.

75 Interview with a key informant in Bungku, 08.09.2012.

76 "Orang Kubu" refers here to the Batin Sembilan.

77 *Daemonorops.*

78 Interview with a key informant in Bungku, 16.09.2012.

79 Interview with a key informant in Bungku, 08.09.2012.
80 Interview with a key informant in Tanjung Lebar, 27.07.2013.
81 Interview with a key informant in Tanjung Lebar, 27.07.2013.
82 Interview with a key informant in Tanjung Lebar, 26.07.2013.
83 Interview with a key informant in Bungku, 08.09.2012.
84 Interview with a key informant in Bungku, 10.09.2012.
85 The land prices are difficult to compare and were paid in different years with different US$-IDR exchange rates. The exchange rate during the last ten years fluctuated between 8455 IDR (2011) and 14120 IDR (2015) for 1 US$.
86 Interview with a key informant in Bungku, 10.09.2012.
87 Interview with a key informant in Bungku, 16.09.2012.
88 Interview with a key informant in Bungku, 12.09.2012.
89 Interview with a key informant in Kampung Laut, 31.08.2013.
90 Interviews with key informants in Kampung Laut, 01.09.2013, 02.09.2013 and in Seponjen, 10.09.2013 and 15.09.2013.
91 Interview with a key informant in Seponjen, 08.09.2013.
92 Interview with a key informant in Bungku, 09.09.2012 and 10.07.2013.
93 Interview with key informants in Bungku, 09.07.2013 and in Tanjung Lebar, 21.07.2013 and 27.07.2013.
94 Fictitious name.
95 Interview with a key informant in Bungku, 10.07.2013.
96 Interview with a key informant in Bungku, 09.09.2012.
97 Interview with a key informant in Bungku, 10.07.2013.
98 Interview with a key informant in Bungku, 10.07.2013.
99 Interview with a key informant in Bungku, 24.08.2013.
100 Interview with key informants in Bungku, 10.07.2013 and 23.08.2013.
101 Interview with a key informant in Bungku, 23.08.2013.
102 Interview with a staff member of Dinas Kehutanan Provinsi Jambi, 19.09.2012.
103 Interview with key informants in Bungku, 14.09.2012 and 09.07.2013.
104 Interview with a key informant in Bungku, conducted by Stefanie Steinebach.
105 Interview with a key informant in Bungku, 09.07.2013.
106 Interview with a key informant in Bungku, 11.09.2012.
107 Interview with a key informant in Bungku, 11.09.2012, 14.09.2012 and 09.07.2013.
108 Interview with key informants in Bungku, 11.09.2012.
109 Interview with a key informant in Bungku, 11.09.2012, 14.09.2012 and 09.07.2013.
110 Interview with a key informant in Tanjung Lebar, 27.07.2013.
111 Interview with a key informant in Tanjung Lebar, 27.07.2013.
112 SAD, refers to Suku Anak Dalam, a post-colonial and deprecatory term for indigenous communities, in this case the interviewee refers to the Batin Sembilan.
113 Interview with a key informant in Tanjung Lebar, 27.07.2013.
114 Fictitious name.
115 Interview with a key informant in Tanjung Lebar, 27.07.2013.
116 Interviews with key informants in Tanjung Lebar, 27.07.2013.
117 Interview with a key informant in Tanjung Lebar, 27.07.2013.
118 Interview with key informants in Tanjung Lebar, 27.07.2013.
119 Interview with a key informant in Tanjung Lebar, 26.07.2013.
120 Interview with key informants in Bungku, 12.09.2012 and 24.08.2013.
121 Interview with an SPI activist in Jambi, 12.07.2013.
122 Interview with an SPI activist in Jambi, 12.07.2013.
123 Interview with key informants in Tanjung Lebar, 25.07.2013 and 27.07.2013.
124 Interview with key informants in Tanjung Lebar, 22.07.2013.
125 Interview with a key informant in Tanjung Lebar, 21.07.2013.
126 Interview with a key informant, in Tanjung Lebar, 21.07.2013.

127 Interview with key informants in Tanjung Lebar, 22.07.2013.
128 Interviews with key informants in Tanjung Lebar, 21.07.2013 and 22.07.2013.
129 Interviews with key informants in Tanjung Lebar, 21.07.2013 and 22.07.2013.
130 Interview with a key informant in Tanjung Lebar, 23.07.2013.
131 Interview with a key informant in Tanjung Lebar, 23.07.2013.
132 The village of Sungai Cemara is not part of the village sample, it is located south of Air Hitam Laut on the coast of the South China Sea. It is very remote and difficult to access.
133 Interview with key informants in Kampung Laut, 29.08.2013 and with a staff member of Dinas Kehutanan Provinsi Jambi, Jambi, 27.08.2013.
134 Interview with a key informant in Kampung Laut, 02.09.2013.
135 Interview with a key informant in Kampung Laut, 29.08.2013.
136 Interviews with key informants in Air Hitam Laut, 29.09.2012, in Kampung Laut, 30.08.2013, 01.09.2013 and 02.09.2013 and in Seponjen, 12.09.2013.
137 Fictitious name.
138 Interview with a key informant in Air Hitam Laut, 29.09.2012.
139 Interview with key informants in Air Hitam Laut, 29.09.2012.
140 Interview with a key informant in Air Hitam Laut, 29.09.2012.
141 Interview with a key informant in Air Hitam Laut, 29.09.2012.
142 Interview with a staff member of Dinas Kehutanan Jambi, Jambi 19.08.2013.
143 Interview with a key informant in Seponjen, 11.09.2013.
144 Interviews with key informants in Seponjen, 11.09.2013 and 15.09.2013.
145 Interview with a key informant in Seponjen, 10.09.2013.
146 Interviews with key informants in Seponjen, 09.09.2013 and 15.09.2013.
147 Interview with a key informant in Seponjen, 10.09.2013.

4   REDD+, privatization and transnationalization of conservation in Indonesia

Private companies and donor governments from the Global North consider forest conservation in the South a cost-efficient option to mitigate climate change. Environmental organizations and public conservation agencies in the South are mainly looking for new options to finance tropical forest conservation. The Indonesian government has initiated a number of governance reforms and established REDD+ pilot provinces in order to be "ready" for foreign investments in carbon conservation. I argue that REDD+ has induced rescaling processes, leading to the emergence of a transnational scale of regulation of forest and land tenure governance. The trading of forest carbon and the allocation of financial contributions from bilateral, multilateral and private donors to governments and forest owners and users requires homogenous rules for land tenure and for the involvement of local communities. Moreover, REDD+ produces a global scale of meaning for local forest conservation efforts since it links place-based forests to the global problem of climate change. It provides entry points for conservationists but also for transnational resistance against offsetting approaches and for climate justice campaigns (Hein et al., 2016).

REDD+ has not only rescaled, it has also transnationalized forest and land tenure governance. The mechanism has strengthened the role of private actors in forest conservation. Private actors have gained influence and authority through implementing conservation projects, and formulating socio-ecological and accounting standards for REDD+ pilot projects and forest carbon offsets. In addition, new forms of hybrid governance (e.g. Pattberg and Stripple, 2008) have emerged where conservation NGOs, private forest carbon standards and companies cooperate in developing rules for and implementing forest conservation initiatives. After years of lobbying and conservationist campaigns led by the Birdlife member Burung Indonesia, since 2008 private actors in Indonesia can apply for conservation concessions within state forest territory.

REDD+ governance and attempts to commodify forest carbon

The growing importance of market-based approaches (such as REDD+) and private actors in conservation can be traced back to the late 1980s and early 1990s. At that time international environmental and development organizations (such as the World Bank) and environmental NGOs increasingly started to argue that environmental problems, such as climate change and forest and ecosystem loss, were caused either by policy failures (Hein, 2013b; McAfee, 1999, 2012a, 2012b) or by a failure to economically account for externalities (Corbera et al., 2009; Hein, 2013b; McAfee, 2012b). To cope with the prevalent environmental crisis in ways compatible with capitalist development, nature or specific eco-system services should be internalized into the economic and financial system (McAfee, 2012b: 26). The idea of conserving and commodifying forest carbon to mitigate climate change came up in the late 1980s when the first private companies and NGOs, mainly from the United States, started to engage in for-est conservation to offset their emissions voluntarily, thereby creating a market for forest carbon credits (Hein, 2014: 508; Hein and Garrelts, 2014: 320; Hein et al., 2015: 2; Neeff et al., 2009: 8). Forest conservation as a potential threat to economic growth has been transformed to a new profit option through the invention of REDD+, carbon trade and payment for ecosystem service schemes (McAfee, 2012b; McGregor, 2010).

The Kyoto Protocol (KP), agreed on in 1997, introduced the Clean Devel-opment Mechanism (CDM) and Joint Implementation (JI) as the first formal emission trading systems permitting the trading of emission rights (Hein, 2014; Hein et al., 2015). The CDM and the JI allow companies to offset their emis-sions, for instance through investing in reforestation and afforestation projects (ibid). Forest conservation and reducing emissions from deforestation were not covered by the KP because of critique raised by the EU and some NGOs regarding the permanence and additionality of climate mitigation based on terrestrial carbon and forest conservation (Bäckstrand and Lövbrand, 2006: 64; Hein et al., 2015: 2).

In 2005, due to strong lobbying by Costa Rica and Papua New Guinea, the issue of deforestation re-entered the UNFCCC negotiations (Hein et al., 2015: 2). Reintroducing forest conservation into the UNFCCC negotiations helped to find a way out of the dead-end of the Kyoto Protocol (Corbera and Schroeder, 2017: 1). Many countries supported REDD+ because forest conservation was perceived as a very cost-efficient option to mitigate climate change (Stern, 2007). Despite all the criticisms, risks and challenges, REDD+ is still perceived as having lower opportunity costs than phasing-out fossil fuel emissions (Angelsen et al., 2014; Houghton et al., 2015). The idea of REDD+ aligns the interests of differ-ent actors in a perfect way, creating momentum for an influential discourse coali-tion involving those interested in forest conservation (e.g. the big environmental NGOs such as Conservation International and The Nature Conservancy) and those interested in developing carbon markets and options to offset greenhouse gas emissions (e.g. companies such as BP and Intel) (Hein et al., 2015: 2).

Fragmented REDD+ governance

The first official decision of the UNFCCC regarding REDD+ was made dur-ing the 13th conference of the parties (COP 13) on the island of Bali, Indo-nesia. The so-called Bali Roadmap includes a decision (Decision 2/CP.13) on

"Reducing emissions from deforestation in developing countries: approaches to stimulate action" (UNFCCC, 2007). The decision encourages developing countries to "[. . .] undertake efforts, including demonstration activities, to address the drivers of deforestation relevant to their national circumstances, with a view to reducing emissions from deforestation and forest degradation and thus enhancing forest carbon stocks due to sustainable management of forests [. . .]" (UNFCCC, 2007). REDD+ gained further political traction after COP 16 in Cancun, Mexico. The Cancun Agreements invite developing countries to formulate national REDD+ strategies and developed countries to support these actions (UNFCCC, 2010). Moreover, the parties agreed on a set of socio-ecological safeguards for implementing forest conservation initiatives.

In 2013, the parties agreed on the Warsaw Framework for REDD+ at COP 19 in Warsaw. The framework consists of seven UNFCCC decisions. They include rules for establishing forest reference levels, monitoring and verification, and criteria for the disbursement of result-based payments (Horstmann and Hein, 2017: 60). The Paris Agreement signed in 2015 refers to forests in article 5. But the agreement does not include additional rules on how to proceed with REDD+ implementation and finance. However, REDD+ plays a prominent role in the National Determined Contributions (NDCs), which are the national climate strategies under the Paris Agreement (Bhan et al., 2017; Corbera and Schroeder, 2017; Hein et al., 2018b). Most of the countries with tropical rainforest cover, including Indonesia, plan to implement REDD+ policies.

In contrast to the CDM or the Green Climate Fund, which are governed by central management bodies, the governance of REDD+ can be considered highly fragmented (Zelli et al., 2014). REDD+ governance consists of a number of decisions by the UNFCCC and of a diverse set of actors. The UNFCCC decisions provide a relatively broad and legally non-binding political framework (Horstmann and Hein, 2017). The diverse set of actors that implement and finance REDD+ pilot projects and larger country-wide REDD+ projects have mostly developed their own rules and financing mechanisms, only guided by the decisions of the UNFCCC. Private actors such as the Climate Change, Community and Biodiversity Alliance (CCBA) and Verified Carbon Standard (VCS) have developed socio-ecological safeguards and accounting standards that facilitate the trade of carbon credits produced through protecting forests (CCBA, 2018; VCS, 2015). In this context a whole new set of actors has emerged such as forest carbon consultants, experts measuring and counting the carbon content of biomass, and conservation companies setting up pro-profit REDD+ projects or using carbon markets to finance existing conservation initiatives. Some of these private REDD+ projects produce carbon credits which can be traded on voluntary carbon markets. Large multilaterals such as the World Bank have established new funding mechanisms like the Forest Carbon Partnership and the Bio Carbon Fund. These are intended to finance forest conservation and support national forest governance reforms to prepare countries for result-based payments and carbon markets. Germany and Norway, for instance, have established their own schemes to finance forest conservation in the Global South. Currently, the International Civil Aviation Organization (ICAO) is considering including REDD+ in its offsetting scheme. This is supported by large environmental NGOs, among others by Conservation International, The Nature Conservancy and the Environmental Defense Fund (Conservation International et al., 2016). The so-called CORSIA (Carbon Offsetting and Reduction Scheme for International Aviation) scheme will also have its own rules and governing bodies (ICAO, 2016).

In a nutshell, REDD+ governance is fragmented and transnationalized. It involves intergovernmental organizations, development banks, national governments and private actors. NGOs, companies and NGO-company hybrids such as the VCS and the Rainforest Alliance increasingly fulfill the traditional role of formal state institutions. They set up conservation rules, provide the regulatory framework for carbon markets, and are involved in rule enforcement (e.g. they conduct audits). Indeed, private actors can withdraw the certification and verification of forest carbon offsets, and – as in the case of the Harapan Rainforest – employ private security to enforce conservation regulations. Furthermore, conservation companies cooperate with governmental institutions in setting up result-based payment initiatives (e.g. VCS cooperates with the Brazilian state of Acre and with a public forest finance fund in Costa Rica (Fondo de Financiamiento Forestal) (Castillo, 2013; Duchelle et al., 2014). In the case of the Berbak Carbon Initiative, the cooperation agreements between the Zoological Society of London and the Provincial Government of Jambi explicitly mention that their project is consistent with VCS and CCBA rules (Dinas Kehutanan Provinsi Jambi, 2013). Thus the provincial government explicitly accepts the regulatory authority of private actors (Hein et al., 2018a: 16).

Contested commodification

REDD+ and forest carbon offsets can be considered as an attempt to assign a price to the ability of forests to store and sequester greenhouse gases (Corbera, 2012). To transform the carbon sequestration services of forests to tradable commodities a number of conditions have to be fulfilled. First, clear property rights to ecosystem services (e.g. the ability of forests and trees to capture greenhouse gases) have to be assigned (privatization) (Castree, 2003: 279–282). Second, ecosystem services have to be exchangeable in order to trade them (alienability) (ibid. 279). Third, ecosystem services should be separated from their supporting context (individuation) (ibid. 280). Fourth, ecosystem services have to be homogenized (abstraction) in order to produce services which are exactly like any other service produced by a forest located elsewhere on the globe (ibid. 281). Fifth, prices have to be assigned (valuation). These rather theoretical steps practically involve several technical steps that are conducted by state agencies (e.g. MoF), carbon consulting companies and forest carbon standard organizations. The starting point is a projection of the amount of greenhouse gases sequestered by the forest ecosystem, including emissions avoided by conserving the forest ecosystem. The amount of sequestered carbon and the emissions avoided by the project determine the amount of tradable emission permits (Hein and Garrelts, 2014: 320). Emission permits or carbon credits then allow the ability of a forest ecosystem to sequester greenhouse gases to be traded. It actually remains unclear whether the implementation of an international REDD+ scheme for financing forest conservation will require the full commodification of forest carbon. To this day, the commodification of forest carbon remains contested – at the scale of UNFCCC negotiations but, as the conflicts described in Chapter 5 illustrate, also at the village and project scale.

It is still unresolved whether forest conservation should be financed through an emission trading system, result-based aid or traditional aid modalities. The Warsaw Framework for REDD+ leaves the question on how to finance REDD+ relatively open, because countries were not able to agree on whether the mechanism should be eligible for offsetting or should be funded through carbon markets or non-market mechanisms and global funds. Consequently, the Warsaw Framework refers to "appropriate market-based approaches and non-market-based approaches [. . .] to support the results-based actions by developing country Parties (UNFCCC, 2013)". Many existing bilateral and multilateral REDD+ funding programs either support forest governance reforms, REDD+ and conservation pilot projects or they pay for achieved emission reductions (e.g. FCPF Carbon Fund). The German REDD+ Early Mover program pays, for instance, USD 5 per tCO2eq (Kreditanstalt für Wiederaufbau, 2017). In Indonesia, the Norwegian Forest and Climate Initiative first paid for governance reforms and later would also pay achieved emission reductions (Government of the Kingdom of Norway and Government of the Republic of Indonesia, 2010). The German International Climate Initiative (IKI), one of the donors of the Harapan Rainforest project, supports actors who manage protected areas. In the case of the Harapan Rainforest, IKI and Danida advertized their financial contribution by referring to the emission-reduction potential of the conservation project, but do not pay directly for quantified emission reductions (DANIDA, 2012a; Internationale Klimaschutzinitiative, 2015).

Even though only few public initiatives directly assign a price to forest carbon, and the market volume of the voluntary carbon markets remains low, REDD+ has changed the conservation logic (Corbera, 2012). The mechanism creates at least the theoretical opportunity to run pro-profit forest conservation initiatives. Even without fully commodifying forest carbon, REDD+ facilitates access to public and private donors for forest owners. Thus, it can foster arguments such as "no pay no care" (Fisher, 2012), but it might also help chronically underfunded conservation authorities in developing countries to receive support from the Global North. A high-ranking official of Jambi's Forest Service argued along these lines and stated, "We want funding for our four national parks. We want compensation for protecting our national parks from international donors. Our national parks are storing CO_2; industrial countries are emitting CO_2 (Hein et al., 2018a)".

"REDD+ safeguards" and the rights of local and indigenous communities

REDD+ safeguards seek to prevent and mitigate potential negative impacts of REDD+ activities. Safeguards are a set of minimal requirements for REDD+ activities. In addition to protecting communities from potential negative effects, the safeguards should provide security for investments in forest conservation by creating homogenous rules for community participation, benefit sharing and environmental integrity. Safeguards have been used by the World Bank and other multilateral development and environmental agencies for decades (Duchelle et al., 2017; Poudyal et al., 2016). Private actors, such as the Plan Vivo and CCBA, already developed safeguards for forest carbon offsets and REDD+ projects in the late 1990s and early 2000s to facilitate the transaction of forest carbon credits (CCBA, 2018; Plan Vivo, 2017).

Safeguards and rights-based language were avoided in early UNFCCC REDD+ negotiations (Jodoin, 2017: 54). However, this changed a few years later at COP 16 in Cancun, Mexico, where parties agreed on a list of social-environmental safeguards. The so-called Cancun Safeguards reflect the ambition to ensure a balance between providing prescriptive rules, accepting national sovereignty and minimizing transaction costs (Jagger et al., 2012: 305). REDD+ activities should "be undertaken in accordance with national development priorities, objectives and circumstances and capabilities and should respect sovereignty" (UNFCCC, 2010). Relevant safeguards referring to the rights of local and indigenous communities are listed in the following (UNFCCC, 2010). REDD+ activities should:

- "be undertaken in accordance with national development priorities, objectives and circumstances and capabilities and should respect sovereignty" (1.e);
- "be consistent with the adaptation needs of the country" (1.h);
- show "respect for the knowledge and rights of indigenous peoples and members of local communities, by taking into account relevant international obligations, national circumstances and laws, and noting that the United Nations General Assembly has adopted the United Nations Declaration on the Rights of Indigenous Peoples" (2.c);
- guarantee "the full and effective participation of relevant stakeholders, in particular indigenous peoples and local communities, in the actions referred to in paragraphs 70 and 72 of this decision"[1] (2.e).

The Cancun Safeguards state that REDD+ interventions should "promote and support" the rights of indigenous communities and "full and effective" stakeholder involvement. However, the formulation of the safeguards is relatively weak, e.g. the decision only "*notes*" that the UN Declaration on the Rights of Indigenous Peoples (UNDRIP) has been adopted. The decision does not stipulate that any REDD+ intervention must follow the declaration and does not mention free, prior and informed consent (FPIC) explicitly.

As an emerging principle in international law, FPIC dates back to Convention 169 of the International Labor Organization (ILO) from 1989, and was picked up by the Convention on Biological Diversity (CBD) in 1992, by the UN Declaration on the Rights of Indigenous Peoples in 2007, and by FAO in 2012. Today many different interpretations of FPIC exist and no general agreement on its meaning and its implications has yet been achieved (Sargent, 2015: 88–89). Generally speaking, communities that are affected by an intervention (e.g. REDD+) that endangers the survival of the community are entitled to FPIC. Especially entitled are indigenous communities or communities that share common characteristics with indigenous communities (UN REDD, 2013: 11–12). In 2013, UN REDD published guidelines for the application of FPIC in the context of REDD+. The guidelines stipulate FPIC for any REDD+ intervention – especially in cases where indigenous populations have to be relocated, propose the mapping of rights before implementation, and recommend the formulation of national FPIC guidelines (UN REDD, 2013: 22–24). Conducting FPIC might be challenging at the local level and may raise the following questions: who is entitled to FPIC and who represents communities in negotiations with implementing agencies? (Hein and Garrelts, 2014: 327). However, it is important to add that the UN REDD guidelines on FPIC are voluntary and do not form part of any formal UNFCCC decision.

Three years after the UNFCCC agreed on the Cancun Safeguards the Warsaw Framework for REDD+ further strengthened the safeguards of the UNFCCC for REDD+ (Horstmann and Hein, 2017). Parties agreed to add that developing parties that seek to receive result-based payments "[. . .] should provide the most recent summary of information on how all of the safeguards referred to in decision 1/CP.16, appendix I, paragraph 2, have been addressed and respected before they can receive result-based payments" (UNFCCC, 2013). Despite the progress in comparison to earlier UNFCCC agreements, many NGOs and climate justice organizations have criticized the Cancun Safeguards as being too weak, too general and for not having clarified the legal status of the safeguards as such (Lang, 2010; Spiller and Fuhr, 2010). However, the safeguards also reflect the relative success of recent indigenous activism in influencing UNFCCC negotiations on REDD+, in contrast to the negotiations on the Kyoto Protocol and Copenhagen Accord where no references to indigenous rights were made (Ciplet, 2014; Jodoin, 2017). At the UNFCCC scale, the adoption of rights-based language by referring to UNDRIP can be considered as an important success for the indigenous rights movement (Ciplet, 2014 and Jodoin, 2017). At the national and regional scale, indigenous rights organizations such as AMAN (Indonesia), AIDESEP (Peru) and COICAA (Amazon Basin) have gained increased attention and have mostly been successful in lobbying for the inclusion of FPIC in national REDD+ safeguards and REDD+ strategies (Ciplet, 2014; Jodoin, 2017; Zelli et al., 2014). In Indonesia AMAN used the slogan "no rights no REDD+" to lobby for the recognition of territorial rights, while the organization also influenced the development of the

PRISAI (Prinsip, Kriteria, Indikator, Safeguard, Indonesia) REDD+ safeguards (UNREDD, 2012: 18).

Despite recent success, REDD+ is still a contested subject within indigenous communities and indigenous rights organizations. In Colombia, for example, the dispute between groups trying to critically engage and "improve" REDD+ and those rejecting the mechanism was openly discussed at a meeting of indigenous leaders of the Colombian Amazon in March 2017.[2] Some participants completely refused attempts to commodify forest carbon, others argued in favor of direct financial benefits for indigenous communities, and a third group tried to use the forum to gain recognition. In contrast, peasant organizations such as the transnational La Via Campesina movement, the Indonesian Peasant Union (SPI), and the Agrarian Reform Movement (Alliansi Gerakan Reforma Agraria, AGRA) strictly oppose REDD+. Moreover, peasants have not been as successful in influencing the UNFCCC negotiations as indigenous groups. The UNFCCC safeguards do not provide much additional recognition for peasants nor do peasants benefit from the inclusion of UNDRIC. In addition, it is worth mentioning that especially in countries that were not impacted by European settler colonialism, such as Indonesia, the entire population can be to some extent be considered indigenous.

Private carbon standards and donor safeguards

Before and after the formulation of safeguards at the scale of UNFCCC, private carbon standards, multilateral REDD+ donors such as UN REDD and bilateral donors such as the German, Danish and Norwegian governments started to develop their own policies to create "no harm". The German bilateral initiatives funded by the BMZ and those funded by Danish bilateral cooperation (DANIDA) are committed to human-rights-based approaches. German and Danish human-rights-based approaches support the UN Declaration on the Rights of Indigenous People, the Convention 169 of the International Labor Organization (ILO) and the implementation of free, prior and informed consent (FPIC) where indigenous communities may be affected (DANIDA, 2011: 3; Schielmann et al., 2013: 26). The Danish government, for example, considers human rights "as means and end in our development cooperation" (DANIDA, 2012b: 2).

The German International Climate Initiative (IKI), which has funded the Harapan Rainforest project, initially had no coherent safeguard policy in place. IKI's funding guidelines only stipulated that implementing organizations have to outline the potential consequences of the project to local and indigenous groups (Schielmann et al., 2013: 27). They have to explain how negative consequences might be avoided or minimized, how the rights of indigenous groups and local communities are respected, how groups are compensated, and how a participatory approach is ensured. The guidelines did not include clear "do no harm guidelines", they did not clarify what is meant by a "participatory approach" and they only stated that existing international standards might be helpful for project implementation (ibid.). PT REKI has been criticized by various actors for neither conducting FPIC nor other forms of participatory community involvement prior to project implementation.[3] Neither were required by IKI. Since 2017 and very likely as a result of the conflicts over the Harapan Rainforest, IKI has incorporated the safeguard approach of the Green Climate Fund (GCF) and the International Finance Cooperation (IFC). Now project implementers have to conduct risk assessments prior to project implementation (Internationale Klimaschutzinitiative, 2017).

REDD+ projects that aim to sell carbon credits on the voluntary carbon markets such as the Berbak Carbon Initiative have to follow certain regulations formulated by transnational carbon standards (e.g. CCBA, VCS, Plan Vivo). Certification according to a carbon standard can be considered a precondition for selling forest carbon credits (Hein and Garrelts, 2014: 321). The Climate Community and Biodiversity Standard (CCBS) was developed by the Climate Community and Biodiversity Alliance (CCBA). CCBA is an organization founded by the NGOs Care, Conservation International, The Nature Conservancy, Rainforest Alliance and the Wildlife Conservation Society, and is supported by the companies BP, Intel, Weyerhauser and GFA Envest (CCBA, 2015). The CCBS provides criteria for assessing REDD+ projects and other forest carbon offsets. Projects certified according to the CCBS should at least create no harm for local communities, and in the best case they should provide benefits for them. Relevant criteria in the second edition of the standard referring to the rights of local and indigenous communities are listed in the following. The project proponents have to:

- "Demonstrate [...] that the project will not encroach uninvited on private property, community property [...] and has obtained the free, prior and informed consent of those whose rights will be affected by the project" (G5.3).
- Demonstrate that the project does not require the involuntary relocation of people or of the activities important for the livelihoods [...]. The project proponents must demonstrate that the agreement was made with the free, prior and informed consent of those concerned and includes provisions for just and fair compensation ("including lands that communities have traditionally owned, occupied or otherwise used or acquired")" (G5.4) (CCBA, 2008 cited in Hein and Garrelts, 2014: 324).

The CCBS stipulates FPIC and recognizes traditionally owned land and land occupied by local communities. CCBS provides far-reaching recognition of local community rights.

Despite, or probably because of, the co-existence of many different approaches towards REDD+ (e.g. UNFCCC, bilateral approaches, voluntary carbon markets), no coherent global or transnational scale of forest governance has yet been formed. The current institutional structure of REDD+ remains fragmented (Zelli et al., 2014: 18). Many different actors, e.g. UNFCCC, donor agencies and non-state actors such as CCBA, have formulated different standards, rules and safeguards and seek to construct their own scales of regulation.

Indonesian REDD+ governance

The highly fragmented institutional structure of REDD+ at the global scale is also reflected at national and sub-national scales (Indrarto et al., 2012). Especially during Indonesia's REDD+ boom period from approximately 2008 to 2012, Indonesia's REDD+ governance was diverse and fragmented involving various sub-national, national, bilateral, multilateral, private and transnational REDD+ initiatives. In 2012, 37 REDD+ pilot projects were initiated (Forest Climate Center, 2012) and public donor initiatives from 11 countries either contributed or committed more than USD 750 million for the implementation of REDD+ policies. Private sector initiatives (including private foundations and NGOs) contribute more than USD 55 million towards REDD+ and good forest governance in Indonesia (Forest Trends, 2014). Over the period 2002–2012, various German donor initiatives contributed USD 33.4 million directly for the forest sector and an additional USD 81.9 million for general environmental protection, including forestry and climate change (Buergin, 2014: 76)

The province of Jambi, for example, simultaneously became a REDD+ pilot province of the former REDD+ Agency, a low carbon-development pilot province of the National Council on Climate Change, and a REDD+ pilot province of the national greenhouse gas reduction strategy (RANGRK). Within Jambi, the provincial REDD+ Taskforce selected specific pilot districts. At the same time, NGOs and conservation companies in cooperation with external donors and state forest agencies established REDD+ pilot projects and landscape-scale conservation initiatives including the Harapan Rainforest and the Berbak Carbon Initiative (Hein, 2013b: 12–13).

National REDD+ governance

The first domestic Indonesian REDD+ policies were outlined in the MoF "REDDI" strategy (REDD Indonesia) shortly before COP 13 in 2007. The document contains first options for emission and deforestation baselines, monitoring of deforestation and benefit sharing (Ministry of Forestry, 2007). In addition, the strategy outlines initial pilot activities to test different approaches that would help to formalize forest governance through gazettement, law enforcement and land tenure assessments (Ministry of Forestry, 2007: 17, 33). In 2008 and 2009 the Ministry of Forestry issued its earliest REDD+ regulations. The first outlined procedures for the implementation of demonstration activities (Regulation 68/2008) and the second outlined procedures for licensing forest carbon projects in production and conservation forest, including rules for benefit sharing between government, communities and the implementing company (Regulation 36/2009) (Menteri Kehutanan, 2008b, 2009). The third regulation (30/2009) further clarifies implementation procedures. Regulation 68/2008 remains very unspecific and does not even mention communities. Regulation 36/2009 states that implementing agencies have to indicate community benefits to receive a license to implement a REDD+ project (article 6) and have to support community empowerment (article 14). Furthermore, the regulation proposes the use of the existing carbon and social-environmental standards of the voluntary carbon market (e.g. CCBS and Plan Vivo) (article 14), strengthening the private governance of REDD+. However, the regulation did not come into force since the Ministry of Finance opposes it.[4] Regulation 30/2009 lists criteria for selecting REDD+ pilot sites e.g. data availability, deforestation risks, and the existence of land conflicts (Ministry of Forestry, 2009). At the project scale in Jambi, the different REDD+ regulations did not seem to be very relevant, only ZSL staff argued that the design of a benefit-sharing mechanism for the Berbak Carbon Initiative is pending because benefit-sharing regulations had not come into force.[5]

In 2009, Indonesia signed the Copenhagen Accord as one of the first emerging economies to announce an emission reduction target. The Indonesian government declared that it would reduce its emissions about 26 percent below business as usual by 2020. Two years later, the former President Susilo Bambang Yudhoyono (SBY) enacted Presidential Regulation No. 61/2011, providing a legal framework for Indonesia's mitigation targets. The regulation includes a detailed work plan for emission reductions, called RANGRK (Rencana Aksi Nasional Pengurangan Emisi Gas Rumah Kaca) and refers to mitigation actions that are – in most instances – part of the existing short- and mid-term national development plans. It is planned that 80 percent of the overall emission reductions target is to be achieved through changes in land use (ibid). The policy proposes emission reductions through the development of plantations on degraded non-forest land, through development of timber plantations, through the issuance of additional private ecosystem restoration concessions, through expansion of community forest concessions and through the development of REDD+ demonstration activities in Jambi and Central Kalimantan (Republic of Indonesia, 2011b). RANGRK consists of the above-mentioned national work plan and provincial work plans (RADGRK), which have to be prepared by the provincial governors (Hein, 2013a: 2).

An influential bilateral agreement that explicitly considers ongoing discussions about land tenure conflicts in the context of REDD+ is the Letter of Intent (LOI) between the Norwegian and Indonesian governments on "Cooperation on reducing greenhouse gas emissions from deforestation and forest degradation". The agreement, signed in 2010, consists of three phases, and foresees payments of up to USD 1 billion in the third phase if Indonesia reaches specific mitigation and policy goals outlined in the LOI (Government of the Kingdom of Norway and Government of the Republic of Indonesia, 2010). The LOI forms the legal basis of the agreement. It calls for "full and effective participation" by all stakeholders, including indigenous groups and local communities, and for the development of "[...] appropriate measures to address land tenure conflicts [...]" (ibid.). The development of appropriate measures to solve land conflicts is listed as one of the official deliverables of the result-based payments agreement, implying at least theoretically that Indonesia will only receive payments from Norway if measures are developed (ibid.). In any case, Norway does not support policy measures aimed at solving land conflicts directly. "The agreement is result based and therefore we do not provide technical assistance for the process" and "[...] we do not use specific social safeguards for the initiative, we draw on Indonesia's domestic safeguards", stated an expert from the Norwegian Forest and Climate Initiative.[6] Other listed deliverables are, inter alia, a two-year moratorium for new conversion concessions within peat and natural forest, the establishment of a special agency directly accountable to the President (REDD+ Agency) and the formulation of a national REDD+ strategy (Government of the Kingdom of Norway and Government of the Republic of Indonesia, 2010).

The moratorium came into force in 2011 and was extended in 2013 and 2015 by the current government under President Jokowi. The moratorium bans new forest and land conversion business licenses (e.g. oil palm concessions) and protects 22.5 million hectares of additional forest and peat land (Murdiyarso et al., 2011). Moreover, the moratorium also fostered the formalization and homogenization of spatial planning processes. The One Map Initiative introduced by the REDD+ Taskforce in 2011 (later upgraded to the REDD+ Agency) to monitor the moratorium is an attempt to develop one reference map for all state agencies (Hein, 2013a). This might reduce the risk of overlapping concessions in the future.

The REDD+ Agency was established in 2013. However, in 2015 the current government closed the agency and integrated REDD+ into the newly formed Ministry for Environment and Forestry (a merger of the Ministry for Environment and the MoF). The national REDD+ Agency was probably an attempt by the Norwegian government to strengthen intersectoral coordination, to circumvent the MoF (Wibowo and Giessen, 2015: 137) and to upscale all REDD+ related policymaking to presidential scale. The agency was a ministerial-level institution established with the aim of supporting the President in coordination, planning and control of REDD+ implementation (Korhonen-Kurki et al., 2017: 67). Consequently, REDD+ was no longer under the authority of the MoF. For some observers, the agency and the national REDD+ strategy led to more transparent and progressive forest governance, e.g. reference to FPIC (ibid.). However, the agency was not successful in implementing policies that reduce deforestation and, in particular, the MoF successfully maintained its strong material power base, namely controlling access to Indonesia's vast state forest territory (Wibowo and Giessen, 2015: 138).

The national REDD+ strategy, published in 2012, was prepared by the REDD+ Taskforce (which was later upgraded to the REDD+ Agency) with the support of civil society organizations. The document summarizes different policies that aim to achieve Indonesia's emission reduction targets and develop Indonesia's forests as a net carbon sink by 2030 (Indonesian REDD+ Task Force, 2012: 5). The Indonesian REDD strategy can be considered as part of

RANGRK or as contributing to RANGRK and at the same time being a deliverable of the agreement with Norway.[7] The strategy entails a framework for the development of social safeguards. Listed safeguards acknowledge the land rights of indigenous and local communities based on historical use (Indonesian REDD+ Task Force, 2012: 28), grant indigenous and local communities the right to reject decisions related to REDD+, and stipulate conflict resolution measures and FPIC. Furthermore, the strategy contains a section on "land tenure reform" that refers to the "[. . .] constitutional right to certainty over boundaries and management rights for natural resources" (ibid. 18), considers land reforms as "[. . .] an important prerequisite to create the conditions required for successful implementation of REDD+", and stipulates FPIC for any new natural resource management concessions (ibid.).

Yet the section on land reform neither explicitly mentions that the state forest will be subject to the reform nor that the MoF will be involved. Only the National Land Agency (Badan Pertanahan Nasional, BPN) and the Ministry of Home Affairs are mentioned as involved actors (Indonesian REDD+ Task Force, 2012: 18). An expert from the MoF argued that "the land reform is meant to further clarify the status of land used by different actors [. . .]" adding that the existing regulations already permit community, smallholder and village forest concessions and implying that the reform will not change the status of the state forest.[8] A CIFOR expert argued, "I don't think that any government will voluntarily conduct a land reform unless there is strong pressure from society".[9] An agrarian reform activist lauded the reference to a land reform in the REDD strategy but added that the REDD+ Taskforce does not have the authority to implement such a reform.[10]

Officially, Indonesia's national REDD+ strategy is still effective but it might be revised by the current government (Kawai et al., 2017: 18). However, Indonesia's National Determined Contribution, the country's national strategy to implement the Paris Agreement, does not mention the REDD+ strategy at all (Hein et al., 2018b). At the end of 2017, the Ministry for Environment and Forestry enacted a new regulation on REDD+. The Ministerial Regulation P. 70/MENLHK/SETJEN/KUM.1/12/2017 outlines Indonesia's pathway to REDD+ implementation. The document repeats that REDD+ activities have to respect the rights of local communities and the right to FPIC of indigenous and local communities. In addition, the document mentions, among other things, result-based payments and carbon markets as potential sources for REDD+ (Menteri Lingkungan Hidup dan Kehutanan, 2017).

The political strategies, policies, agreements and regulations introduced here can be considered as an attempt (not successful) by specific apparatuses of the state to circumvent the MoF and establish new national scales of regulation for implementing REDD+. The political momentum for forest reform provided by REDD+ was used to initiate the formalization of forest governance and spatial planning and to commodify forest carbon. At the same time, indigenous rights organizations tried to benefit from the momentum. REDD+ and the high level of attention directed towards Indonesia's forests may have positively influenced the constitutional court ruling on the release of *adat* forest from the state forest, as an AMAN activist argued prior to the court decision (see Chapter 5 for more detailed information on the constitutional court ruling).[11] Furthermore, AMAN was able to map territories claimed by indigenous groups in cooperation with the REDD+ Agency.

Yet the legal character of the new REDD+ related scales of regulation remains weak. It is, for instance, contested whether disregarding the presidential instruction, which is the legal foundation for the moratorium, has legal consequences or not (Hein, 2013a: 4; Murdiyarso et al., 2011: 2). Furthermore, the legal basis of the LOI and of the National REDD+ strategy is challenged by some actors. A CIFOR expert argued that, "[. . .] in the National REDD+ strategy, there is no explanation and no reference to a presidential decree or regulation".[12] Above all, REDD+ related scales of meaning and regulation are challenged by other far more influential policies such as the Master Plan for Acceleration and Expansion of Economic Development 2011–2025 (MP3EI), which refers to the expansion of mining and corporate oil palm and rubber plantations. Sumatra, for instance, is designated for oil palm, rubber and coal mining (Republic of Indonesia, 2011a: 49). According to the plan, the provinces of Jambi, North Sumatra and Riau should become major "oil palm plantation nodes" (ibid. 51). Central and East Kalimantan are designated as "timber activity nodes". REDD+ is only mentioned as a potential source of non-timber related income. Moreover, timber plantations and not natural forests are considered for absorbing additional carbon under an international REDD+ scheme (ibid. 112). Under the new President Jokowi, REDD+ governance was re-integrated into the formal forest apparatus. For some observers, this indicates that climate change mitigation then became less important for the Indonesian government. The current mid-term development plan (2015–2019), for example, does not refer to REDD+ at all (Korhonen-Kurki et al., 2017: 69).

Jambi's provincial REDD+ governance

Jambi was one of the first provinces in Indonesia that hosted conservation projects linking local conservation efforts with climate change mitigation. The "Climate Change, Forest and Peatlands in Indonesia Berbak-Sembilang Project" (CCFPI), 2002–2005) in Jambi and South Sumatra explicitly linked forest and peat land conservation with mitigation and carbon sequestration objectives (Lubis and Suryadiputra, 2004: 115). The project was funded by the Canadian Climate Change Development Fund. The initiative can be considered as a precursor of the Berbak Carbon Initiative. The first province-wide mitigation strategy with a strong focus on land-based activities (e.g. REDD+ and reforestation) was developed by the National Council on Climate Change and by the Provincial Government of Jambi (Dewan Nasional Perubahan Iklim, 2010; Purnomo et al., 2012: 75).

In 2010, the provincial government started its first attempt to become one of Indonesia's REDD+ pilot provinces. The government of Jambi prepared a draft

REDD+ strategy. The document "Jambi Sebagai Provinsi Percontohan Untuk Mekansime REDD+" (Jambi as a pilot province for the REDD+ mechanism) was developed to outline Jambi's potential as a national REDD+ pilot province (Hein, 2013b: 12; Pemerintah Provinsi Jambi, 2011). The strategy supports the designation of new community forest concessions (e.g. village forests and community forests) in Jambi and argues for an acknowledgement of indigenous and local community rights to forest land. At that time, Jambi's attempt to become a REDD+ pilot province was not successful. Central Kalimantan became the first pilot province.

In 2011, the province started its second attempt and established the Jambi Regional Commission for REDD+ (KOMDA REDD+) through Governor Decree No. 356/2011 (Hein, 2013b: 12). The members of the commission were appointed by the provincial government. In 2013, the commission consisted of the NGOs WARSI, ZSL and Flora and Fauna International; of the conservation company PT REKI; of representatives of the provincial planning agency (BAPPEDA), the provincial forest agency and of the provincial environmental protection authority; and of experts from the University of Jambi and the World Agroforestry Center (ICRAF). The commission has no regulatory authority and was mainly formed to coordinate drawing up the provincial REDD+ strategy. In 2013, Jambi finally became a REDD+ pilot province of the National REDD Agency.

The provincial REDD+ strategy covers the period from 2012 to 2030. In addition to providing technical details and data on the emission-reduction potential of specific land-use policies, the strategy refers to the potential pro-poor benefits of REDD. The strategy also aims to strengthen the rights of local communities and includes plans to map forest land claimed by local communities and indigenous groups (Perbatakusuma et al., 2012). The strategy was written by the commission. Other NGOs were invited to focus-group discussions to comment on the document. The strategy designates the districts of Tebo, Muaro Jambi and Merangin as REDD+ pilot districts. The pilot districts were selected because they represent three different landscapes. Merangin is characterized by the upper Sungai Batang Asai watershed and the Bukit Barisan Range. Whereas Tebo is characterized by hilly low lands and low mountain ranges, Muaro Jambi consists of hilly low lands, tidal marshlands and peat swamps. In addition, the districts were selected because of their high numbers of land conflicts.[13] In these districts, land-use and forest-use permit procedures are to be reviewed, and forest monitoring and law enforcement are to be improved. However, among village and sub-district governments and administrations knowledge of the strategy and the proposed activities seemed to be limited in 2013. The village government of Tanjung Lebar and staff of the administration of the sub-district of Kumpeh in Muaro Jambi, for example, were aware neither of being located within a REDD+ pilot district nor of REDD+ related policy shifts at the district level.[14]

Furthermore, the provincial REDD+ strategy argues in favor of strengthening the rights of local communities and indigenous communities. However, many of the suggested policies cannot be implemented by provincial or district governments alone since they involve national scale authorities such as the MoEF. Furthermore, the provincial government of Jambi has recently acted in ways contradictive to the policies suggested in the REDD+ strategies. According to regional media, the Governor of Jambi refused to sign recommendations for nineteen village forest concessions in the province (JambiekspressNews, 2013). However, it is important to add that Jambi, with its 54,978 ha of village forest concessions managed by 25 villages, has the largest village forest area of all the provinces in Indonesia (Bakhori, 2013). Through different strategies and policies, the provincial government has successfully constructed REDD+ as a new scale of meaning in order to attract donor investments, but it lacks the authority and probably also the political will to establish a complementary scale of regulation.

REDD+ in Jambi: an actor mapping

In this section, I will briefly present a mapping of actors involved either in the implementation of REDD+, or in resistance campaigns against REDD+ in Jambi. The mapping (Figure 4.1) is mainly based on expert knowledge and shows formal state actors and private actors directly involved in REDD+, conservation projects and organizations supporting peasant resistance. Stakeholders from state and non-state organizations were asked to identify actors they consider relevant in relation to the implementation of REDD+ in Jambi. In addition, the mapping and the selection of actors is based on document review and on my own judgment.[15]

The most important state actors are, as a matter of course, the various forest authorities, more explicitly the MoF (MoEF) and the different district and provincial forest services. The MoF has authority over the state forest, has enacted various REDD+ regulations and has allocated the Meranti-River-Kapuas-River forest block to the conservation company PT REKI for implementation of the Harapan Rainforest project. A second relevant national-scale authority was the REDD+ Agency (Badan Pengelola REDD+) which has recently been integrated into the MoEF. The REDD+ Agency was directly accountable to the president. Furthermore, the agency appointed Jambi as an official REDD+ pilot province. The different district and provincial forest services are involved in managing forest reserves and conservation forests. They do not have the authority to allocate concessions but they are involved in conflict mediation, forest monitoring and law enforcement. At the provincial level, the KOMDA REDD+ (Provincial REDD+ commission, Komisi Daerah REDD+) is in charge of coordinating REDD+ implementation. However, as the national REDD+ agency, KOMDA REDD+ has no regulatory power. KOMDA REDD+ is a hybrid institution consisting of state and non-state actors. ZSL, the NGO leading the Berbak Carbon project, is a member of KOMDA REDD+.

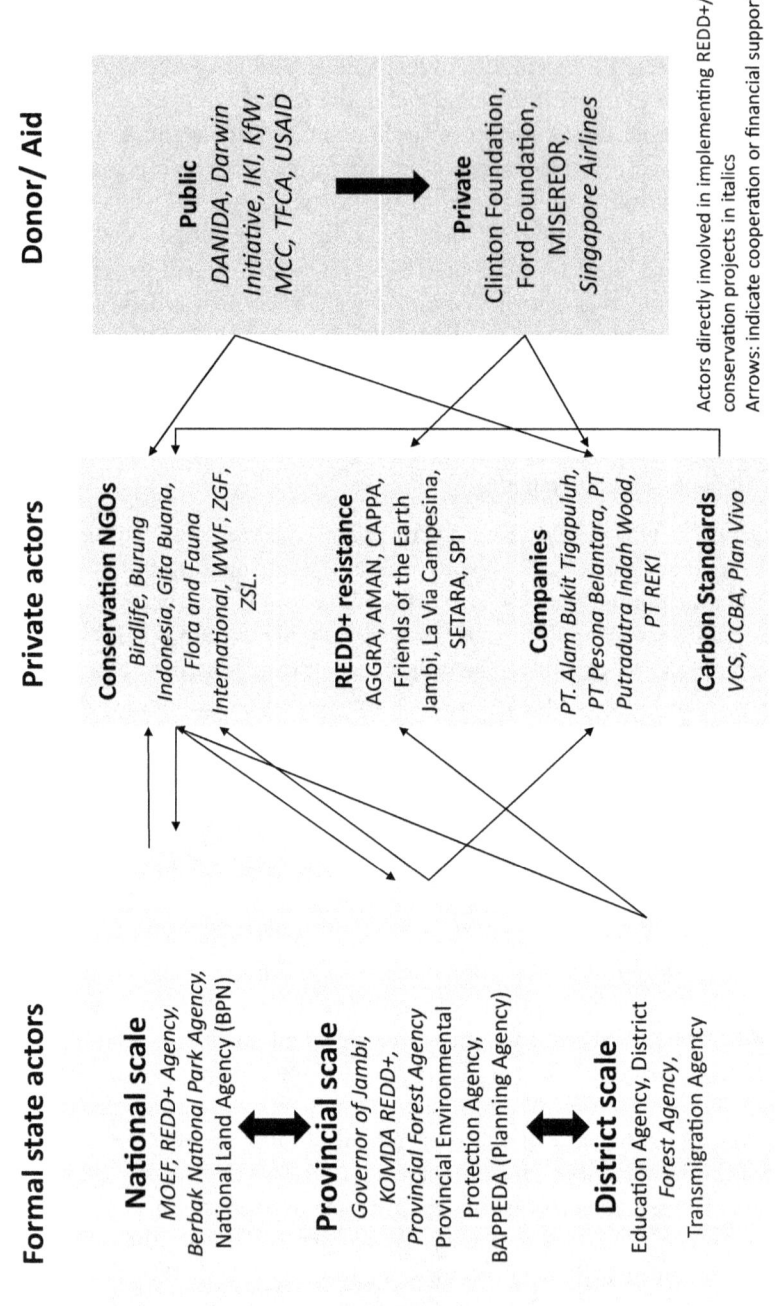

Figure 4.1 Actors involved in REDD+ and conservation in the Province of Jambi

(Source: based on author's own investigations)

State actors challenging the implementation of REDD+ and conservation projects and the integrity of the state forest are the various sectoral agencies of the district governments of Muaro Jambi and Batang Hari. By providing support to peasants or by providing public infrastructure they legitimize the presence of peasants and indigenous groups in the state forest territory. The Transmigration Agency of Muaro Jambi, for instance, has developed transmigration settlements within the borders of the Berbak Carbon project (see Chapter 5 for more details). The Education Agencies (Dinas Pendidikan) of both districts support and run schools within the Harapan Rainforest project. The Agricultural Agency of Batang Hari has provided extension services for settlers farming within the Harapan Rainforest.

The ZSL and Gita Buana are among the important non-state actors involved in the Berbak Carbon project. ZSL identifies itself as an international scientific, conservation and educational charity (ZSL, 2015) and is based in the United Kingdom (UK). The organization started its Indonesia field program in 2002, focusing on ecological research, core area protection and the connectivity of conservation areas (CIRAD, 2012). On the island of Sumatra, ZSL's activities focus mainly on tiger conservation. ZSL seeks to develop mechanisms permitting rural development and conservation. Gita Buana is a Jambi-based NGO founded in 2001 by students from the Faculty of Agriculture at the University of Jambi. Gita Buana is involved in the implementation of FPIC for the Berbak Carbon Initiative. Gita Buana's donors are the Clinton Foundation, Tropical Forest Conservation Action (TFCA) and Flora and Fauna International.[16]

Non-state actors are of key importance in the Harapan Rainforest project. PT REKI is owned by three NGOs. Burung Indonesia was founded in 2002 and focuses mainly on bird and bird habitat conservation. Burung Indonesia works mainly in Sumatra, Sumba, Gorontalo, Sangihe Taluad, Tanimbar and Buru. Burung Indonesia is a member of Birdlife International. The Royal Society for the Protection of Birds (RSPB), also member of Birdlife International, was founded in 1889 in the UK. The RSPB holds more than 200 bird reserves (RSPB, 2015). In the accounting year 2014/2015, the organization had a net income of £99 million (ibid.). Birdlife International is the transnational umbrella association of 120 bird conservation organizations and claims to be the world's largest conservation initiative and the world's leader in bird conservation (Birdlife International, 2015).

Peasants and indigenous groups living within the borders of the Harapan Rainforest project are supported by a number of non-state actors as well. The most relevant are Serikat Petani Indonesia (Indonesian Peasant Union, SPI), and the grass roots organizations Yayasan CAPPA and Yayasan Setara. SPI is the largest Indonesian Peasant Movement and a member of La Via Campesina. La Via Campesina is an umbrella organization uniting 164 local and national peasant organizations from 73 countries. La Via Campesina claims to represent 200 million farmers from all continents. The main aims of the organization are the realization of food sovereignty and the implementation of agrarian reforms.

Yayasan CAPPA is a Jambi-based environmental justice organization.[17] The organization first mainly supported peasants in conflicts with pulp and paper companies. Today the organization claims to support the "victims of agrarian policies". Yayasan CAPPA helps peasants defend and access property rights and argues for social transformation in order to end agrarian conflicts. Furthermore, Yayasan CAPPA is involved in conflict mediation, advocacy work and indigenous rights issues. Its main donors are Misereor, the Ford Foundation and USAID. Yayasan SETARA is a Jambi-based NGO working on environmental justice issues, especially on the consequences of oil palm expansion.[18] Yayasan SETARA supports small-scale oil palm cultivators and indigenous communities. SETARA is a member of the Roundtable for Sustainable Palm Oil (RSPO). The organization has a watchdog function within RSPO and strives for the acknowledgement of peasant rights. Yayasan SETARA and CAPPA cooperate intensively and support Batin Sembilan communities in land conflicts with the conservation company PT REKI and with the oil palm company PT Asiatic Persada. SETARA receives financial support from Misereor.

The mapping indicates linkages (across scales) between formal state actors and linkages among private actors and between state and private actors. Actors promoting REDD+ and conservation and actors resisting the expansion of conservation have established actor coalitions involving local, national and transnational NGOs (see Chapter 5 for more detail).

Privatization and transnationalization of conservation: conservation concessions and co-management

As a biodiversity hotspot, Indonesia has long been a key country for the projects and campaigns of transnational conservation organizations. Transnational NGOs initiated activities in Indonesia in the late 1980s (McCarthy, 2006: 183; Peluso et al., 2008: 383). In those days, the New Order regime of President Suharto took up the emerging sustainability discourse, before and after the first Rio Summit, in order to improve its international reputation (Hall et al., 2011: 68). Conservation NGOs successfully lobbied for new protected areas, national parks and environmentally friendly natural resource management regulations, and engaged in the co-management of conservation areas and community-based conservation projects (Peluso et al., 2008; Wells et al., 1999).

In the 1990s, transnational NGOs like World Wildlife Federation (WWF) and The Nature Conservancy (TNC) co-managed Integrated Conservation and Development Projects (ICDPs) with state authorities and local communities. The ICDP concept aims to go beyond fortress conservation by actively involving local communities and represented the state of the art of conservation at that time (Wells et al., 1999: 13). In 1992, for example, TNC started to support the National Park Authorities of Lore Lindu National Park in Central Sulawesi. The NGO negotiated conservation agreements with communities and established a stakeholder platform to facilitate dialogue among local communities and the different institutions working within and around the national park (The Nature Conservancy, n.d.; Wells et al., 1999: 26). However, the project was not able to deal with the land disputes and land occupations of local and indigenous communities inside the national park (Adiwibowo, 2005). In Jambi, the WWF lobbied for a ICDP project in the buffer zone of Bukit Tiga Belas National Park (Wells et al., 1999: 27). The MoF rejected the plan. Nevertheless, more than 20 years later the WWF was able to realize its plans. Recently, the buffer zone became an ecosystem restoration concession managed by the conservation company PT Alam Bukit Tigapuluh, which was co-founded by the German NGO Zoological Society Frankfurt (ZGF) and WWF Indonesia.

In contrast to Latin American countries, it was relatively difficult for NGOs and companies to establish private protected areas in Indonesia until the 2000s as the Forest Law did not permit a "non-productive use" of forest concessions. In the 2000s, the idea of "governance through markets" (Peet et al., 2011: 7) gained further political traction. In line with the neoliberal zeitgeist, deforestation and biodiversity loss were primarily considered as governance failure and as a failure to economically account for ecosystem services in an adequate manner. The integration of conservation and ecosystem services into the market system would create economic incentives for conservation and would consequently almost automatically lead to better protection. In this context, large environmental NGOs became increasingly powerful. Globally, companies and public donors transferred millions of dollars to conservation organizations, expecting that privately managed conservation would be more cost-efficient than state-managed approaches which were often considered prone to corruption (Igoe et al., 2010). Conservation NGOs now command over 50 percent of the global funding for conservation (Chapin 2004 cited in Igoe et al., 2010: 491) and, in addition to their successful lobbing for new state-managed protected areas, have established an increasing number of privately managed protected areas. In 2015, Birdlife International and its members, for example, had a joint budget of USD 539 million (Birdlife international, 2015). Their 1553 protected areas cover 4.3 million hectares, a territory larger than Switzerland.

In 2004, a forest management reform in Indonesia made the country more attractive for private investment in conservation. The MoF introduced a new concession type (officially: ecosystem restoration concessions), delegating for the very first time authority over conservation, forest rehabilitation and REDD+ activities within the state forest territory to non-state actors (Walsh et al., 2012a: 35). Conservation and ecosystem restoration were integrated into Indonesia's market-oriented land-use planning system (Radjawali et al., 2017), and thus they had to compete with other possible land uses, such as logging and pulp and paper production. According to Rice (2002), the conservation concession concept provides a number of benefits: "Under a conservation concession, conservation becomes a product that can be purchased directly and provided according to clearly established criteria. In combination with payments, the limited term of a conservation concession makes it an attractive option for resource owners. At the same time, long-term conservation is possible because of renewable terms, low opportunity cost, and high willingness and ability to pay" (Rice, 2002: 4). Additional momentum for the conservation concession concept and for private conservation actors in Indonesia can be traced back to COP 13 in Bali. After the Indonesian COP, the role of non-state and state actors in forest conservation changed fundamentally both in Indonesia and globally. The hope that global carbon markets might create a reliable funding source for protected areas or even a source for profit created a short Indonesian "REDD+ rush" (McGregor, 2010). NGOs and corporate actors started to search for potential sites for REDD+ projects; they developed rules and certification mechanisms and projects (Hein and Garrelts, 2014; Newell and Paterson, 2010). Even investment banks like Merril Lynch, the German insurer Allianz, and the oil and gas companies Shell and Gazprom started to invest in forest conservation (Hein and Garrelts, 2014).

In Indonesia, the formulation and adoption of the conservation concession concept was very much the result of strong lobbying by the Indonesian branch of BirdLife International (Burung Indonesia) (Hein, 2013b). In 2007, PT REKI received the first ecosystem restoration concession for the southern part of the Harapan Rainforest project. In 2017, 16 concessions existed all over Indonesia covering 623,075 ha of land. The MoEF announced that it plans to allocate an additional 100,000 ha per year to conservation companies until 2019 (Ministry of Environment and Forestry, 2017). The allocation of a conservation concession to a concession holder goes hand in hand with the transfer of additional bundles of rights replacing former state functions. Environmental protection, protected area management, environmental monitoring and even the allocation of land-use permits for communities within the concession area are de facto exercised by the concession holder.

However, the plans to allocate conservation concessions are challenged by a lack of areas designated for ecosystem restoration, the requirement to obtain a letter of recommendation from the local governments, and a lack of guidance about what governors should recommend and the role of district governments (Walsh et al., 2012: 14). District governments often have little interest in having an ecosystem restoration concession within their jurisdiction. An expert from the MoF argued, "Ecosystem restoration in the short-term is a sacrifice [. . .] for all actors at all levels, including the company".[19] As ecosystem restoration concessions contribute less to the district's economy than other options, some districts seek to impede their issuance.

A further push towards privatization and the commodification of conservation and ecosystem services might be induced by a new regulation on economic instruments for conservation (Government Regulation No. 46 of 2017). The policy foresees compensation payments for the provision of environmental services, including carbon capture and sequestration, and stipulates that polluters should compensate for ecosystem losses. This might also generate additional income sources for actors holding ecosystem restoration concessions.

The private actors involved in the two conservation projects covered by this book used two different approaches. PT REKI and the shareholders Burung Indonesia, the Royal Society for the Protection of Birds (RSPB), actively lobbied for forest management reform and later on received the first conservation concession. In contrast, the approach of the Zoological Society London (ZSL) builds on a more traditional co-management approach.

Lobbying for and accessing conservation concessions[20]

In the early 2000s, Burung Indonesia and its national and transnational partners discussed new conservation and forest rehabilitation concepts, especially for areas designated as production forests. The RPSB was searching for areas to be used as privately financed and managed conservation areas. A key informant from Burung Indonesia explained, "At that time, the UK community over there, the members of RSPB, they wanted to share, for example, if you give money you can conserve 1 hectare".[21] At that time, conservation and forest rehabilitation activities were limited to protected areas managed by state actors. Burung Indonesia argued that conservation regulations for protected areas were too restrictive for private initiatives, and developed a proposal for a new concession type that would permit private actors to invest in conservation. According to a key informant of Burung Indonesia, at first the Ministry of Forestry (MoF) was skeptical, but after years of lobbying, it finally agreed: "Our CEO at that time, he was very persistent, [. . .] he lobbied here and there, also he wanted me to lobby here and there [. . .] I think the third minister finally agreed".[22] In 2004, the MoF enacted the first regulation on "Ecosystem Restoration in Production Forest" (SK 159/Menhut-II/2004) that permitted private conservation concessions. Three years later, the Indonesian parliament integrated private conservation concessions into the Forest Law (Birdlife International, 2008: 21).

An expert from the consulting company Carbon Synthesis, who supported the lobbying campaign, argued that Burung Indonesia's campaign was successful because of the reduced demand for logging concessions that resulted from the post-Suharto boom in illegal logging.[23] The MoF considered the new concession type mainly as a mechanism for recovering state forest production capacity.[24] Burung Indonesia and its national and transnational partners considered the new concession type a response to the failure of state-led conservation policies and an opportunity to finance conservation activities through sustainable forest exploitation, ecotourism and carbon trade.[25]

The Harapan Rainforest

In parallel to Burung Indonesia's lobbying campaigns, in 2000 the NGO started a site selection process with its partners. Biodiversity, urgency, administrative and socio-economic criteria were used to study 14 sites across Indonesia (Birdlife International, 2008: 13). Burung Indonesia conducted the first on-site investigations in Jambi and South Sumatra provinces between 2000 and 2004.[26] The Meranti River-Kapuas River forest block[27] in Jambi and South Sumatra was selected because it represents one of Sumatra's last patches of intact lowland rainforest. The area was also considered to be of great interest to timber and oil palm companies – and consequently at risk (ibid. 13). At that time, the southern part of the concession consisted of a retired logging concession formerly used by the state-owned company, PT Inhutani V. Although the northern part was still owned by PT Asialog, the company had stopped logging in 1999, abandoned the concession in 2003, and signed an agreement with Burung Indonesia about officially ending logging activities in the concession in 2006 (Birdlife International, 2008: 22; Mardiana, 2014: 14). In 2005, the MoF officially classified the forest block for ecosystem restoration.[28]

Initially, Jambi's local governments opposed the permit for the Harapan Rainforest project.[29] One year later, the NGO consortium won the public tender for the southern part of the concession and received letters of support from the district heads and governors of Jambi and South Sumatra (Birdlife International, 2008: 22). The MoF officially revoked PT Asialog's logging concession in 2007 (Mardiana, 2014: 14). In 2008, PT REKI, the company founded by the NGO consortium to manage the concession, received the southern part and two years later the northern part of the Meranti-River-Kapuas-River forest block. The concessions are valid for 100 years. The permit process for the conservation concession took eight years (Birdlife International, 2008: 22).[30]

The MoF neglected the presence of the settlements (Figure 3.1) of local communities and indigenous groups (settlememt formation described in Chapter 3) and allocated major parts of the Meranti-River-Kapuas-River forest block to the conservation company. In the following years a number of land conflicts emerged (some of which continue), challenging project implementation and the ability of PT REKI to control its territory (Chapter 5).

The Harapan Rainforest is a privately managed project focusing on biodiversity conservation and climate change mitigation (Figure 4.2). Still, large shares of the funding for the project come from public sources. The German International Climate Initiative (IKI) and the Danish Ministry of Foreign Affairs (DANIDA) are probably the most important public donors. IKI provided EUR 7.5 million from the end of 2009 to 2013. PT REKI has applied for a second funding period but a decision was still pending in early 2018. In 2016, staff from the German Embassy in Jakarta seemed to be in favor of a second period, explaining that not continuing to support the first privately managed conservation project in Indonesia would be a negative political signal. The German embassy considers the project a role model for conservation in Indonesia. From 2011 to 2016 DANIDA provided EUR 9 million through its Special Climate Change Fund's Fast-Start Finance (FSF) (Buergin, 2014: 65). Additional funds of DKK 20 million were provided in the framework of the Danish-Indonesian Environment Support Program (DANIDA, 2016). The most important private donor is Singapore Airlines. The carrier contributed US$ 3 million and became the "exclusive airline partner" of the project. The carrier considers the project as part of its strategy to improve its overall environmental performance (Singapore Airlines, 2015).

Figure 4.2 Picture of destroyed forest and border sign of the Harapan Rainforest
(Source: taken by the author, 2013)

The Indonesian REDD+ Agency listed the Harapan Rainforest as a REDD+ demonstration project (Badan Pengelola REDD+, 2014). The current project list from the new MoEF no longer lists the project. The question as to whether the Harapan Rainforest is a REDD+ project seemed to be an ongoing dispute among different shareholders of the conservation company. PT REKI seems to dissociate itself more and more from REDD+. In October 2013, a project manager from Burung Indonesia stated that "[...] we were from the very beginning not a REDD+ project, PT REKI is not REDD [...]".[31] But in 2012 a project manager based in Jambi argued that the "[...] carbon market is a long-term funding option, the baseline preparation is in process and we will base a carbon consultant in Palembang" (in South Sumatra).[32] PT REKI's main donors IKI, DANIDA and Singapore Airlines explicitly name climate mitigation, carbon sequestration or REDD+ piloting as the project objectives (DANIDA, 2012a; Internationale Klimaschutzinitiative, 2015; Singapore Airlines, 2015). PT REKI's recent attempts to disassociate itself from REDD+ might be considered as strategy to avoid potential controversies on benefit sharing and the commodification of forest carbon or as response to SPI's climate justice and anti-REDD+ campaigns (Hein and Faust, 2014: 24). So far PT REKI has not sold any carbon credits and has not been certified according to a carbon standard.

Conservation agreements and benefits of the Harapan Rainforest

The implementation of conservation regulations and community benefits is challenged by ongoing conflicts over access to land. According to interviewed PT REKI staff, the company initially planned to develop two zones with specific access rights for peasants and for the Batin Sembilan. The first, called the *tanaman kehidupan zone*, was to provide the Batin Sembilan with opportunities to gather non-timber forest products and plant rubber. The second *mitra zone* should provide land for peasant migrants who were living in the concession before the Harapan Rainforest project started. Initially, PT REKI only accepted the presence of peasant migrants who were living in the settlement before 2010 (when the project was initiated in Jambi). A staff member of the conservation company explicitly stated, "Those that came after have to leave. We want to push them out".[33] Migrants living in the concession before 2010 should pay rent to PT REKI and have to follow existing regulations for state forest land, e.g. implying that oil palm cultivation and shifting cultivation are prohibited. PT REKI proposed a maximum plot size of 2–5 ha per migrant household for the cultivation of forest crops like rubber.[34]

In addition, PT REKI negotiated and is still negotiating a number of conservation agreements (Village Resources Management Agreement, VRMA) with the indigenous Batin Sembilan, and recently with some peasant migrants. The agreements usually allow the communities to use a certain amount of land for rubber and fruit tree cultivation and to harvest non-timber forest products, but prohibit land trades and additional forest conversion (REKI, 2011b) (see Chapter 5 for additional information on conservation agreements in the context of conflict mediation and solution).

As part of the conservation agreements (VRMA) and beyond, PT REKI provides a number of direct income-related and non-income-related benefits (Table 4.1). The benefits are intended to provide alternative sources of income, stabilize household incomes, and compensate the opportunity costs of PT REKI's intervention. Non-income-related benefits provided by PT REKI include elementary school services, free health care, clean water, electricity and better sanitation. However, these benefits are de facto only accessible for the Batin-Sembilan-dominated settlements of Simpang Macan (Dalam and Luar) and the Batin Sembilan living in the *mitra* settlement.[35] Direct income-related benefits provided by PT REKI include employment opportunities, agricultural extension services, and the joint marketing of non-timber forest products. PT REKI provides employment opportunities for local communities living in the Harapan Rainforest, the most important is the community nurseries. Community nurseries have been established in the Batin-Sembilan-dominated settlements of Simpang Macan (Dalam and Luar, part of Kunangan Jaya I, Bungku), in the *mitra* (part of Kunangan Jaya I, Bungku), and in Sako Suban. Each community nursery provides employment for approximately seven families. The staff of the nurseries are paid for each seedling, for the planting of the seedling at the reforestation site, and if the seedling survives the first year. The nurseries

Table 4.1 Conditional land tenure and community benefits of the Berbak Carbon and the Harapan Rainforest Projects

Project	Important regulations	Conditional land tenure / conservation agreements	Other community benefits	
			Income-related	*Non-income-related*
Harapan Rainforest	Only forest species (trees), food-crop cultivation is permitted*, no commercial logging, no land transactions	1–5 ha per household (allocation and negotiations ongoing)	Employment (e.g. in community nurseries and as forest rangers), agricultural extension services, marketing	Health services, education, sanitation, electricity
Berbak Carbon Initiative	Only forest species (trees), logging is prohibited, no land transactions	1–5 ha per household (first plots allocated in Seponjen and Kampung Laut)	Agricultural extension services, employment; further benefits depend on governmental regulation	Planned

(Source: compiled by the author)

* At least in the TSM settlement.

provide monthly incomes of between IDR 300,000 and 700,000 per per-son.[36] Community nurseries have not yet been established in the settlements described in Chapter 3. Thus, peasant migrants have de facto no access to this source of alternative income. Inhabitants of the *mitra* settlement have the additional opportunity to earn monthly payments of IDR 500,000 for providing information about incidents such as forest fires or for the denunciation of new so-called "encroachers".

PT REKI staff members claimed that if possible, full-time positions would be filled with local community members. Nevertheless, in July 2013, only two inhabitants of the *mitra* settlement had permanent jobs – as PT REKI forest guards.[37] PT REKI's reforestation efforts provide additional short-term employment opportunities. For example, PT REKI pays groups of planters up to IDR 1,600,000 per ha.[38] Agricultural extension services provided by PT REKI consist mainly of the provision of rubber seeds and livestock.

An additional benefit that PT REKI aims to provide especially for Batin Sembilan is support for the marketing of forest products. PT REKI already organizes the marketing of honey for Batin Sembilan (Wardah, 2013: 19) and plans to support the marketing of jelutung and jernang (dragon blood[39]).

PT REKI has developed new conservation regulations, tree inventories, and a preliminary carbon assessment and has started border demarcation. Moreover, PT REKI has established a forest guard unit to protect the conservation concession and monitor compliance with the conservation regulations. PT REKI's activities can be considered an active attempt to construct a new scale of meaning – linking local conservation efforts to global environmental problems such as biodiversity loss and climate change – and to establish the project area as a new scale of regulation. As we will see in Chapter 5, this scale of regulation is highly contested.

The Berbak Carbon Initiative and ZSL's co-management approach

The ZSL's approach to establishing the Berbak Carbon Initiative differs substantially from that of Burung Indonesia and its partners. The ZSL's basic idea is to harmonize the management of various concessions and protected areas – to establish a new scale of regulation for the Berbak landscape. ZSL has not tried to gain full ownership or full control over its different land-use categories. ZSL's strategy is to influence the management of existing concessions and protected areas through negotiating cooperation agreements with concession owners and forest authorities. A key informant working for ZSL argued that applying for a private conservation concession is not an option for the NGO because of the high upfront costs. He said, "It is too expensive for NGOs with no money. Because we have to have a very big investment [. . .] to get the concession area. Yeah, that's also, I think, the obstacle for an NGO who wants to get the restoration concession for REDD activity. Big investment [. . .] while to get the restoration area required the tax for the first 35 years, plus an extension of

60 years: 65 years must be paid in advance".[40] In addition to the financial barriers, only the logging concessions managed by PT Putra Dutra Indahwood and PT Persona Belantara Persada would be eligible for reclassification as private conservation concessions. State forest that is already classified as a protected area, such as the Berbak National Park, is not eligible to become a private conservation concession.

The Berbak Carbon Initiative was also listed as a REDD+ demonstration project by the Indonesian REDD+ Agency. The project is a collaborative initiative by the Zoological Society of London (ZSL), the Jambi-based NGO Gita Buana, the Berbak National Park Agency and the Provincial Forest Service (Dinas Kehutanan Provinsi Jambi), and can be considered as the latest attempt to establish the Berbak (carbon) landscape as a scale of meaning and regulation. The project has to cope with informal forest conversion activities and with a transmigration settlement established by the Transmigration Agency of the district of Muaro Jambi. The transmigration settlement was constructed within the state forest and within the borders of the Berbak Carbon Initiative, violating MoF regulations.

The first discussions about establishing a REDD+ project between the ZSL and the Berbak National Park Agency started in 2008. The first cooperation agreement on establishing a REDD+ project in the Berbak landscape was signed with the MoF in May 2011, and with the Berbak National Park Agency in October 2011. The agreement with the National Park Agency defines various areas of cooperation including the establishment of a measurement, reporting and verification system, support of the VCS and CCBS certification processes and the development of community benefits (Balai Taman Nasional Berbak and Zoological Society London, 2011). The document states that potential income from carbon trade will be used to preserve biodiversity and maintain the peat-swamp forest's carbon storage capacity (ibid.). In 2013, the ZSL signed a third agreement with the Provincial Forest Agency on implementing a REDD+ initiative in the Sektitar Tanjung forest reserve. The agreement defines, inter alia, FPIC implementation, the closure of illegal drainage canals, and the development of a benefit-sharing mechanism for carbon trade revenues (Dinas Kehutanan Provinsi Jambi, 2013).

Initial funding of USD 478,883 (The REDD Desk, 2015) for the project was provided by the Darwin Initiative of the UK Department for Food, Environment and Rural Affairs (DEFRA). The carbon credit broker Eco Securities announced its interest in purchasing credits from the project in 2008. Yet, to date, the project has not been certified and has not sold any carbon credits. ZSL is seeking for CCBS and Verified Carbon Standard (VCS) certification for the project. The Indonesian-based carbon consultancy company Forest Carbon has conducted an "initial field and desktop assessment of carbon emission reduction potential for the Berbak Carbon Initiative" (Eickhoff et al., 2010). The authors conclude that the initiative could lead to emission reductions of approximately 75–82 MtCO2e and is eligible for certification according to the VCS standard (ibid. 7). The project is of specific interest since it is planned to be implemented in different forest categories (Berbak National Park, Forest Reserve Sekitar Tanjung, limited production forest (logging concessions) and conservation forest), permitting different types of land use. However, at the time of field research only the Berbak National Park Agency and the Provincial Forest Service managing the Forest Reserve had officially agreed to form part of the REDD+ initiative. Negotiations with the logging companies PT Putra Dutra Indahwood and PT Persona Belantara Persada were not finalized at this time. According to the ZSL, the companies were mainly concerned about the great uncertainties regarding carbon trade revenues.[41] After years of delay, ZSL announced in 2015 that the project will be operational in 2018 (ZSL, 2015). However, the project is still not listed in the VCS database.

Existing forest regulations are an additional obstacle to harmonizing forest management in the landscapes of the Berbak Carbon Initiative. The forest reserve and national park had to be registered as separate REDD+ projects since it was not possible to register them as a single REDD+ project under current forest regulations.[42] The ZSL sought to circumvent the regulatory barriers by negotiating different cooperation agreements but has not yet been able to establish a coherent new scale of regulation. The various concessions and conservation areas are still governed by different authorities. A staff member of the Provincial Forest Agency argued that coordination was difficult since no authority wanted to lose influence to another.[43] Jumping to the transnational scale would still allow for the construction of a new conservation scale of regulation for the Berbak area. Certification under VCS and CCBS would create a set of rules applicable to all land-use categories. So far, the ZSL has only managed to establish REDD+ as a new scale of meaning for the landscape. Other actors (e.g. logging companies) are challenging the production of a complementary scale of regulation.

Implementing the Berbak Carbon Initiative: conditional land tenure and FPIC

Potential benefits for local communities from carbon trade profits were still under discussion. An ZSL project manager stated that benefit sharing is planned but its design will depend on national legislation.[44] A first community-needs assessment for designing community benefits was conducted by the NGO WALESTRA (Wahana Pelestarian dan Advokasi Hutan Sumatera) and the first pilot community reforestation and conditional land-tenure schemes are running in Seponjen, Kampung Laut and a few other villages in the Berbak landscape.

In contrast to PT REKI and as requested by the CCBS carbon standard, ZSL and the project partners have started to conduct a formalized free prior and informed consent (FPIC) process in 32 villages around the Berbak Carbon Initiative. In 2013, the process was underway in 11 villages, including Seponjen and Kampung Laut. However, the first consultation sessions in 2013 only involved members of the village elite and were conducted in the office of the sub-district head (*camat*) – not in the villages. In the case of Seponjen, only the village head, the imam, the head of the village parliament, and the neighborhood heads participated.[45] The village population was not officially informed by its representatives about the outcome of the consultation meeting. In Kampung Laut and Seponjen, therefore, knowledge about the Berbak Carbon Initiative and REDD+ in general was relatively limited. In Kampung Laut, only three of 17 community members interviewed had heard of REDD+. A member of the village government argued that he has not disseminated information on REDD+ since he does not want to create high expectations in REDD+ as a potential income source.[46]

A local leader who had participated in the meetings with Gita Buana, the NGO conducting the sessions, stated that they were informed about REDD+, especially that "the world will buy the carbon that is stored in our region".[47] In a progress report, Gita Buana noted that the community representatives recognized the importance of the Berbak ecosystem and agreed to contribute to its protection (Gita Buana, 2013). In addition to the FPIC process, a number of different community consultation sessions were conducted in the three study villages – mostly to provide information about regulations for the protected areas and the conditional land tenure and community reforestation program.

Because most forests in the project area have been protected for many years, project implementation has led to only limited changes. Furthermore, only a few households use agricultural land within the boundaries of the Berbak Carbon Initiative. Like the Harapan Rainforest, the Berbak Carbon Initiative has de jure not established any additional land-use restrictions, although it may have contributed to better enforcement and better acceptance of the existing legal framework for forest land. Additional regulations associated with the project tend rather to promote community involvement, such as FPIC implementation, than to further restrict access to land. Important land-use restrictions mentioned by key village informants were the prohibitions on oil palm cultivation, annual crops, and agricultural activities in the buffer zone (within 500 m of the project boundary), logging, and open fires.[48]

The main benefits provided by the various stakeholders implementing the Berbak Carbon Initiative are conditional land-tenure schemes[49] (Table 4.1) and employment opportunities in the community reforestation programs. Benefits for emission reductions are dependent on the planned benefit-sharing regulations for forest carbon projects, which are conditional on selling carbon credits. The ZSL and its partners have conducted a community-needs assessment to design benefits for the community.[50]

The conditional land tenure and community reforestation scheme in the forest reserve is funded by the MoF's reforestation fund and through the provincial budget and backed by Law No. 5/1990 and Governmental Regulation No. 38/2007 (Republic Indonesia, 1990, 2007). The scheme aims to allocate land to farming groups in eight villages around the Berbak Carbon Initiative and will privilege poor households and households that traditionally own land in the project area. The allocated land rights are based on memorandums of understanding (MoU) between the farming groups and the Provincial Forest Service. In Seponjen, the Provincial Forest Service has designated 150 ha of the forest reserve for farming groups. The farming groups have up to 25 members and receive 2 ha of land per participant; this might be expanded to 5 ha in the future. Trading land is prohibited: the land can only be used for jelutung,[51] rubber, and rambutan[52] cultivation.[53]

In September 2013, 20 farming group members in Seponjen received 1 ha of land and were paid IDR 120,000 a day to prepare and plant the land. This is the first group to be officially registered by the Provincial Forest Service. According to a farming group member, not all interested community members had received land: many are on a waiting list. He explained that the households that are "ready" and those who need land had been selected first, but he added that the village government and the forest service had not used clear criteria to select participants.[54] Another farming group member stated that many community members were unaware that they could receive land free of charge from the forest service.[55] The information provided by key informants in Kampung Laut was partly contradictive. One key informant stated that land allocation had not yet started in Kampung Laut and explained that households interested in receiving land should form farming groups and prepare management plan proposals.[56] A second key informant said that he knew about the program but thought that it had not started yet.[57] A third key informant who is a member of a farming group said that the program was already running: his farming group, which already had 37 members and an official permit from the Provincial Forest Service, had already started to plant jelutung trees.[58] A fourth informant stated that 50 ha of land would be allocated to farmers in Kampung Laut: the mapping and allocation of 2-ha plots had just started.[59] When visiting the project area in August 2016, most of plots that had been allocated to farmers had been destroyed by the devastating peat and forest fires of that year.

The non-transparent dissemination of information on the conditional land-tenure scheme and knowledge asymmetries between different community members permitted elite capture and rent-seeking behavior. An example from an area of the forest reserve claimed by both the village of Seponjen and the sub-district capital of Tanjung/Suakandis is illustrative. A migrant from Lampung stated that he had bought land that was designated for the conditional land-tenure scheme – from a field assistant of the Provincial Forest Service. He paid IDR 7,000,000 for 3 ha of land, including a *sporadik title* issued by the head of the Kelurahan Tanjung/Suakandis. In all, 30 ha of land designated for the scheme had been sold to ten households.[60]

Land conflicts associated with the Berbak Carbon Project are not as severe and violent as in the Harapan Rainforest project. Forest authorities tolerate fishing and jelutung tapping in the National Park. Land conversion within the project occurs, e.g. east of Seponjen, and, as mentioned, initiated by the construction of the transmigration settlement in Kampung Laut.

Other private and/or donor-funded REDD+ activities in Jambi

Private actors have gained importance in Jambi as their strong involvement in the Harapan Rainforest and Berbak Carbon Initiative illustrate. However, these are only two of a number of new private conservation initiatives in the province (Table 4.2). Many of them are again privately managed but funded by public donors.

Table 4.2 REDD+ demonstration projects in Jambi[61]

Project	Ecosystem	Project status	Land-use category/ concession	Proponents	Funding	FPIC
Harapan Rainforest	Dry lowland rainforest	Running, no tradable certificates yet, pre-carbon accounting done	Limited production forest/ERC–Concession	Burung Indonesia, Birdlife International, RSPB, PT REKI	KfW until 2013, Danida, Harapan Fund (e.g. Singapore Airlines), Conservation International	No formal FPIC
Berbak Carbon Initiative	Peat-swamp forest, mangroves	Running, no tradable certificates yet. Verified Carbon Standard (VCS) and CCBS in preparation	National park, conservation forest, forest reserve and limited production forest	ZSL, Gita Buana, National Park Agency, Province of Jambi, PT Pesona Belantara Persada, PT Putraduta IndahWood	Darwin Initiative, Clinton Foundation, UK Panthera Fund	Ongoing
Bukit Tigapuluh Ecosystem Conservation	Dry lowland rainforest and mountain rainforest	Running, pre-carbon accounting done	Production forest/ ERC–Concession	ZGF, WWF, Yayasan Kehati, The Orang Utan Project, National Park Agency, PT Alam Bukit Tigapuluh	KfW, TFCA	Controversies on community consultation
Village Forest Community Carbon Pool	Mountain rainforest	Project is fully operational, approved by PlanVivo carbon standard	Production forest, village forest concession	Flora and Fauna International (FFI), Village Forest Management Institution	Darwin Initiative, International Carbon Action Partnership	Yes according to FFI
Community Forest Management Project in Jangkat Highland	Mountain rainforest	Project Idea Note approved by the Plan Vivo carbon standard	Production forest, village forest concession	SSS Pundi Sumatera, Village Forest Managing Agency	No information	No information

(Source: Badan Pengelola REDD+, 2014; Forest Climate Center, 2012; Hein, 2013b; own interviews)

The **Bukit Tigapuluh Ecosystem Conservation initiative** was developed by the Frankfurt Zoological Society (FZS), the Bukit Tigapuluh National Park Agency, WWF, Yayasan Kehati and other NGOs. The NGOs ZGF, WWF and Yayasan Kehati have recently founded a conservation company (PT. Alam Bukit Tigapuluh) and received a private conservation concession (ERC) in 2015. The project receives EUR 3.6 million from IKI through KfW to build up the project infrastructure (ZGF, 2016). As in other projects, the involved NGOs plan community benefit schemes to improve the livelihoods of local communities. Initially, ZGF planned to finance future project activities via carbon trade and recently conducted a carbon pre-assessment (Hein 2013b: 14). Parts of the project area are occupied by local indigenous groups and by Javanese migrants. Friends of the Earth Jambi (WALHI Jambi) issued a press statement in February 2016 in the name of the village community of Pemayungan. The community complained that villagers have to stop farming inside the concession and that they were not informed prior to project implementation (WALHI Jambi, 2016).

The **Community Forests for Climate, People and Wildlife** project is an initiative of Flora and Fauna International (FFI) with project sites in Jambi and West Kalimantan (Fauna & Flora International, 2012). FFI plans to support villages by implementing voluntary market REDD+ projects in their village forest concessions. In Jambi FFI collaborated with the village community of Durian Rambun in the Merangin district. The village holds a village forest concession of 4484 ha (ibid. 4). The project is operational and was validated in 2015 and registered in 2017 (Plan Vivo, 2018).

The **Community Forest Management Project in Jangkat Highland** is implemented by the NGO Sumatra Sustainable Support Pundi (SSS Pundi Sumatra) in the villages of Pematang Pauh, Talang Tembago and Muara Madras in the Merangin district (Pundi Sumatera, 2014: 2). The Project Idea Note was approved by Plan Vivo in 2014 (Plan Vivo, 2015).

In addition, Jambi is a target area of the Tropical Forest Conservation Action (TFCA) for Sumatra. TFCA is a US-Indonesian debt-for-nature swap administered by the Indonesian NGO Yayasan Kehati (TFCA-Sumatera, 2014). Its main focuses are biodiversity conservation at the landscape level and contributing to Indonesia's emission reduction targets. TFCA currently supports village border demarcation in buffer zone villages of the Berbak Carbon Initiative. TFCA also provides funding for the Bukit Tigapuluh Ecosystem Conservation Initiative. A second major donor initiative linking conservation and mitigation is funded through the US Millennium Challenge Cooperation (MCC). MCC supports land-use planning and alternative livelihood strategies in 32 villages surrounding the Berbak Carbon Initiative (Millennium Challenge Account, 2015: 20).

Summary and preliminary conclusion

REDD+, market-based conservation and general discursive shifts associated with neoliberal conservation (Fletcher, 2010) and ideas to sell nature to save it

(McAfee, 1999) have changed Indonesia's conservation landscape. Conservation NGOs have become increasingly influential. Until the early 2000s, they were mainly involved in lobbying for new protected areas, the co-management of protected areas and campaigning for stricter environmental regulations. Today, conservation NGOs have established private protected areas and conservation companies in order to access conservation concessions and to trade with ecosystem services. Moreover, they are involved in formulating and enforcing socio-ecological safeguards and carbon certification systems. NGOs and conservation companies have taken on the former functions of state agencies: they develop rules for managing protected areas, trade with ecosystem services and have established private security agencies to protect their conservancies.

REDD+ in particular has created a whole new set of (mostly non-binding) rules, recommendations and guidelines. The trading of REDD+ credits and multilateral or bilateral result-based payments for emission reductions require homogenous rules for carbon accounting but also to govern the acknowledgement of community rights and participation processes. The result is an emerging transnational scale of REDD+ governance. However, REDD+ has not yet led to coherent and uniform transnational rules. The current situation is rather characterized by a number of fragmented, sometimes competing scales of regulation. This is also reflected at the national and sub-national scales where many different forest-based mitigation initiatives exist in parallel to each other, some of which even compete or contradict one another. Even more importantly, many of the new Indonesian REDD+ regulations, policies, strategies and letters of intent have not yet been implemented or have been challenged by other state apparatuses, hindering the construction of a coherent national scale of REDD+ regulation. Furthermore, many of the MoF REDD+ regulations remain unspecific and seem to be irrelevant at the project scale. Recently, the closing of the REDD+ Agency, the lack of reference to REDD+ in Indonesia's NDC and in a recent mid-term development plan, indicate that REDD+ is a contested mechanism at the national scale.

The various UNFCCC legal documents, the criteria of transnational carbon standards (e.g. CCBS), and the safeguards of donors (e.g. the BMZ human rights policy) have at least potentially strengthened the rights of local and especially of indigenous communities vis-à-vis national forest agencies, private companies, conservation NGOs and donors. In this context, REDD+ may have changed the dialectical relationship between structure and agency. Indications for this changed relationship are AMAN's success at the constitutional court, the consideration of FPIC in forest carbon standards, the acknowledgements of customary land tenure in Indonesia's national REDD+ strategy, the Cancun Safeguards, and probably also the ability of SPI to use the Harapan Rainforest project for SPI's and La Via Campesina's transnational campaigns against offsetting.

REDD+ and forest carbon offsetting link emitters in the North to land conflicts in the Global South, thus contributing to the transnationalization of alleged local land-tenure conflicts. REDD+ provides entry points for transnational resistance campaigns tackling questions of global climate justice. The following Chapter 5 focuses on this new kind of transnationalized agrarian conflict that arises in the context of the implementation of privatized conservation and REDD+ initiatives in Indonesia. The simplistic logic of market environmentalism meets the complexity of Jambi's land tenure regime outlined in Chapter 3. In this situation, new conflicts and new alliances among actors are emerging while old pre-existing conflicts change their meanings.

Notes

1 Article 72 requests developing country parties "[…] when developing and implementing their national strategies or action plans, to address, inter alia, the drivers of deforestation and forest degradation, land tenure issues, forest governance issues, gender considerations and the safeguards identified in paragraph 2 of appendix I to this decision, ensuring the full and effective participation of relevant stakeholders, inter alia indigenous peoples and local communities".

2 This meeting was held in Bogota in the context of the German-, Norwegian- and UK-financed REDD+ early mover program. I had the opportunity to participate.

3 Interviews with key informants in Bungku, 12.09.2012 and in Tanjung Leber, 21.07.2013 and with CAPPA activists in Jambi, 18.07.2013 and based on an unpublished KfW document.

4 Interviews with a staff member of the Ministry of Finance, Jakarta, 19.07.2012 and with a staff member of CIFOR, Bogor, 26.07.2012.

5 Interview with a staff member of ZSL, Bogor, 27.08.2012.

6 Interview with a staff member of the Norwegian Forest and Climate Initiative, via telephone, 11.12.2012.

7 Interview with a staff member of CIFOR, Bogor, 26.07.2012 and with a staff member of the Ministry for Finance, Jakarta, 19.07.2012.

8 Interview with a staff member of the Ministry of Forestry, Jakarta, 23.07.2012.

9 Interview with a staff member of CIFOR, Bogor, 26.07.2012.

10 Interview with a staff member of Sajogyo Institute, Bogor, 13.08.2013.

11 Interview with an AMAN activist, Jakarta, 27.07.2012.

12 Interview with a staff member of CIFOR, Bogor, 26.07.2012.

13 Interview with a member of the KOMDA REDD+, Jambi, 15.08.2013.

14 Interview with a staff member of the sub-district administration of Kumpeh, Tanjung/Suakandis, 02.09.2013, and with a key informant in Tanjung Lebar, 25.07.2013.

15 Here, I do not describe all actors mentioned in Figure 4.1 in detail. I only describe the role of the most relevant actors; others were introduced earlier or will be introduced in Chapter 5.

16 Interview with activists of Gita Buana, 22.08.2013, Jambi.

17 Interview with activists of CAPPA, 18.07.2013, Jambi.

18 Interview with activists of SETARA, 18.07.2013, Jambi.

19 Interview with a staff member of the Ministry of Forestry, Jakarta, 23.07.2012.

20 The terms ecosystem restoration concession and private conservation concession are used as synonyms.

21 Interview with a staff member of Burung Indonesia, Bogor, 11.10.2012.

22 Interview with a staff member of Burung Indonesia, Bogor, 11.10.2012.

23 Interview with a staff member of Carbon Synthesis, Jakarta, 11.10.2012.

24 Interview with a staff member of the Ministry of Forestry, Jakarta 19.07.2012 and 23.07.2012.

25 Interview with a staff member of Carbon Synthesis, Jakarta, 11.10.2012.
26 Interview with a staff member of Burung Indonesia, Bogor, 11.10.2013.
27 Official name used by the Ministry of Forestry for the production forest block located on both sides of the border region of Jambi and South Sumatra.
28 Decree of the Minister of Forestry No. SK.83/Menhut-II/2005, dated April 1, 2005.
29 Interview with a staff member of Burung Indonesia, Bogor, 11.10.2012.
30 Interview with a staff member of Burung Indonesia, Bogor, 11.10.2012.
31 Interview with a staff member of Burung Indonesia, 11.10.2013, Bogor.
32 Interview with a staff member of PT REKI, 02.09.2012, Jambi.
33 Interview with a PT REKI staff member in Bungku, 30.07.2013.
34 Interviews with a PT REKI staff member in Jambi, 02.09.2012 and with a PT REKI staff member in Bungku, 23.09.2012, 30.07.2013 and 31.07.2013.
35 The *mitra* settlement is a resettlement initiative for Batin Sembilan of the conservation company PT REKI. The participating families lived before scattered over the concession area.
36 Interviews with a PT REKI staff member in Jambi, 02.09.2012 and with a PT REKI staff member in Bungku, 31.07.2013.
37 Interview with a PT REKI staff member in Bungku, 31.07.2013.
38 Interview with a key informant in Bungku, 12.09.2012.
39 Dragon blood, Daemonorops draco, is a specific rattan species that produces a red resin.
40 Interview with an ZSL staff member, Bogor, 27.08.2012.
41 Interviews with an ZSL staff member, Bogor, 27.08.2012 and staff of the Berbak National Park Agency, Jambi, 19.08.2013.
42 Interview with an ZSL staff member, Bogor, 27.08.2012.
43 Interview with a staff member of Dinas Kehutanan Provinsi Jambi, 27.08.2013.
44 Interview with an ZSL staff member, Bogor, 27.08.2012.
45 Interviews with key informants in Seponjen, 10.09.2013 and 11.09.2013.
46 Interview with a key informant, Kampung Laut, 30.08.2013.
47 Interview with a key informant in Kampung Laut, 30.08.2013.
48 Interview with key informants in Seponjen, 08.09.2013, 12.09.2013 and 13.09.2013 and in Kampung Laut, 29.08.2012 and 04.09.2013.
49 Various actors run the different community involvement programs, the most advanced of which is the community reforestation and conditional land-tenure program run by Jambi's provincial forest service.
50 Interview with a staff member of the ZSL in Bogor, 27.08.2012.
51 *Dyera costulata*, syn. *D. laxiflora*
52 *Nephelium lappaceum*
53 Interview with a staff member of Dinas Kehutanan Jambi in Jambi, 27.08.2013.
54 Interview with a key informant in Seponjen, 09.09. 2013.
55 Interview with a key informant in Seponjen, 10.09.2013.
56 Interview with a key informant in Kampung Laut, 30.09.2013.
57 Interview with a key informant in Kampung Laut, 01.09.2013.
58 Interview with a key informant in Kampung Laut, 29.08.2013.
59 Interview with a key informant in Kampung Laut, 28.08.2013.
60 Interview with a key informant in Seponjen, 12.09.2013.
61 Interviews with a project manager of ZGF, Jambi, 31.08.2012, with staff of PT REKI, Jambi, 02.09.2012 and with a staff member of Burung Indonesia, 11.10.2013.

5 Transnationalized agrarian conflicts in the REDD+

This chapter focuses on the transnational dimensions of agrarian conflicts and peasant resistance. It investigates the different resistance strategies of peasants and indigenous groups vis-à-vis conservation NGOs and apparatuses of the state. It provides successful examples of peasant resistance challenging the commodification of forest carbon and disentangles conflicts between development and conservation-oriented apparatuses of the state.

Access to and control of Sumatra's forests have been subject to contestation, especially after the regime change of the late 1990s. As outlined in Chapter 3, peasants and indigenous communities supported by sub-national governments were able to challenge centralized authority over the state forest territory and started to regain control of former customary land. However, the state forest territory remained under the de jure control of the Ministry of Forestry, leading to highly complex and ambiguous land and forest tenure and to sometimes violent conflict involving peasants, indigenous groups, various state agencies and corporate actors. Conflicts among peasants and oil palm and timber plantation companies were notably violent. However, they also fostered the formation of local resistance movements and provided the opportunity for peasants to establish alliances with the transnational and national peasant and indigenous rights organizations that emerged across the archipelago after the regime change.

This dynamic and contested political landscape provides context for REDD+ and market-based conservation interventions and peasant resistance on the island of Sumatra. In particular, REDD+ and the increasing influence of private actors described in Chapter 4 have not only changed Indonesia's conservation landscape, they have also changed the meanings and the spatial extent of apparently local conflicts about access to and control over forests. REDD+ has strong implications for environmental and climate justice. It links actors with emission-intensive lifestyles in the Global North and in urban centers with peasants and indigenous groups living at the forest margins, leadings to new transnational alliances and new transnational conflicts. Apparently local conflicts about Jambi's forests became conflicts about the ability to define a forest as a "carbon toilet for the rich countries" or as a resource for regional development, swidden farming and agricultural expansion.

REDD+ and the emergence of the indigenous rights discourse changed the dialectical relationships between structure and agency. The acknowledgement of indigenous rights in international law as well as in the emerging REDD+ governance framework provided additional legitimacy for ethnicity-based land claims and entry points for transnational NGOs. REDD+ has increased international attention towards structural inequality in national forest governance in the tropics and has fostered the diffusion of international norms such as FPIC, strengthening the position of peasants and indigenous groups vis-à-vis forest authorities, companies and conservation project developers. While the debates on REDD+ have widened the room for maneuver for some peasants and indigenous communities, the simultaneous expansion of conservation areas and large-scale agricultural estates has limited land-use opportunities for most peasants and indigenous communities significantly (Hall, Hirsch and Li, 2011 Hein and Faust, 2014; Hein and Garrelts, 2014; Kijazi, 2015; Osborne, 2011).

The conflicts that occurred in the context of the implementation of the Harapan Rainforest and the Berbak Carbon Initiative involve peasants, indigenous groups, different state agencies and – in the case of the Harapan Rainforest – the conservation company PT REKI. The cases illustrate a transnationalization of agrarian conflicts. Moreover, the conflicts described in this chapter can be considered as attempts by peasants and indigenous groups (and also by sub-national governments) to defend the access to and control over forest and resources they had at least partially gained in the turbulent early post-Suharto period (described in Chapter 3). The various village-scale settlements located within the state forest territory described in Chapter 3 are now challenged by REDD+ and conservation interventions (Figure 5.1). Not all state agencies involved in

Figure 5.1 Picture of entrance portal to a settlement of Kunangan Jaya I located inside the Harapan Rainforest

(Source: taken by the author, 2013)

local contestations over land act in line with national policies, some have actively supported peasants in conflicts, some sub-national state agencies have engaged in open resistance against centralized control of the state forest territory.

This chapter provides, first, an introduction to the indigenous rights movement and agrarian movements and their attempts to challenge centralized control of state forest territory and to lobby for agrarian reform and indigenous rights. Second, it gives a short overview of conflicts among peasants, indigenous groups and oil palm companies, and of the implications of these conflicts for land struggles in the context of REDD+. Third, it provides in-depth analysis of specific multi-scalar and transnational land conflicts characterized by different conflict histories, scalar arrangements and actor constellations. Peasants and indigenous groups employ multiple strategies to legitimize and maintain their presence within the state forest territory and within areas designated for forest conservation and REDD+.

The formation of resistance movements and alternative scales of meaning and regulation

Public invasion of the state forest territory in Jambi and other parts of Indonesia was actively supported by peasant and indigenous rights organizations, in particular after the fall of Suharto, but invasions also took place in the context of earlier regime changes (e.g. in the early post-colonial period). After the fall of Suharto, peasants and indigenous groups started to establish initial factions in order to fight against historically contingent structural marginalization and violence (Peluso et al., 2008: 379). Today, indigenous rights and peasant organizations are powerful mass organizations with well-established transnational support networks. The most important organizations are probably Aliansi Masyarakat Adat Nusantara (AMAN, Indigenous Rights Movement of the Archipelago) and the Serikat Petani Indonesia (SPI, Indonesian Peasant Union). Whereas SPI is directly involved in the struggles over access to and control of the Harapan Rainforest, AMAN is rather indirectly involved through lobbying for indigenous rights to land and self-determination. Other important national peasant and indigenous rights organizations involved in agrarian conflicts in Jambi are Serikat Tani Nasional (STN, National Peasant Union) and Alliansi Gerakan Reforma Agraria (AGRA, Alliance Movement for Agrarian Reform). In addition, a number of local organizations and advocacy groups exist in Jambi. Many of them were founded in the context of ongoing land conflicts with oil palm and pulp and paper companies. Organizations involved in the conflicts described in this chapter include the community-based organization SAD 113 (indigenous groups of the three villages) and advocacy NGOs such as CAPPA and SETARA (c.f. chapter 4).

Indigenous rights organizations and peasant organizations cooperate, but their narratives and frames are based on fundamentally different ideological foundations. AMAN, as the largest group of the indigenous rights movement, clearly links land rights to ethnic identity. Agrarian organizations such as SPI link land rights to citizenship, to the Indonesian constitution and to rights guaranteed in the Basic Agrarian Law (Peluso et al., 2008: 387; Rachman, 2011: 104; Tuong, 2009: 183). Consequently, both movements refer to different scales, narratives and regulations. AMAN seeks to reestablish different local and ethnic-specific scales of cultural meaning and scales of land tenure regulation based on pre-existing ethnic territories. Agrarian organizations fight for a more socially and environmentally just national scale of regulation. In practice, despite the ideological differences, the two movements cooperate. For instance, AMAN and SPI jointly called on members to vote for President Jokowi in the 2014 presidential elections (Alliansi Masyarakat Adat Nusantara (AMAN) Bengkulu, 2014).

The reemergence of adat and the foundation of AMAN

Indigenous communities in Indonesia, in general known as *adat* (customary) communities, regained influence during the *Reformasi* era. *Adat* communities and their leaders increasingly started to claim land based on pre-existing ancestral lands and pre-existing larger *adat* territories (e.g. *Wilayah Adat*) (Barr et al., 2006: 2; McCarthy, 2005: 57). Members of the indigenous Batin Sembilan living within and adjacent to the Harapan Rainforest project, for instance, claim that almost the entire project area used to be part of their former customary land. Pre-existing *adat* territories and customary law were also used by district governments to legitimate the allocation of small-scale timber and forest conversion concessions (Barr et al., 2006: 12; Rhee, 2009: 46). In many areas, a parallel village administration based on customary law appeared. Furthermore, ethnicity became a relevant category for the legitimation of political authority (Rhee, 2009: 46). Categories such as *putra daerah* (literally, child of the region) became relevant and conflicts spread over the archipelago, especially between migrant groups and the indigenous population (Acciaioli, 2001: 87; Rhee, 2009: 46). In Jambi, interviewed villagers reported only smaller ethnic disputes, and interethnic marriages are common.

Especially the occupation of state forest or regaining access to state forest was a means to overcome the historical structural injustice and political marginalization that started with the Dutch conquest and was aggravated during the New Order era (Barr et al., 2006: 2; Peluso et al., 2008: 386; Steinebach, 2013b: 65). Furthermore, decentralization and shifting boundaries caused by jurisdictional reforms revealed the way in which the scales of regulation established by the Dutch colonial government and the Suharto government did not necessarily fit with pre-existing settlement patterns, traditional authority and customary territories.

The reemergence of *adat* or, in broader terms, of ethnicity as a means to claim natural resources was not accidental. Decentralization laws such as Law 22/1999 explicitly permitted more diverse local administration that considered local customs (Moeliono and Dermawan, 2006: 115). In 2001, the Indonesian parliament decided that land tenure and natural resource management laws have to "[...] recognize, respect and protect the rights of adat law communities"

(Indonesian Supreme Parliament (MPR), cited in McCarthy, 2005:58). Further-more, the strong transnational indigenous rights movement, as well as interna-tional agreements and conventions such as Convention 169 of the International Labor Organization (ILO), the Convention on Biological Diversity (CBD) and the UN Declaration on the Rights of Indigenous People, provided additional agency and significantly enhanced the political context of *adat* groups in Indo-nesia (Bedner and Van Huis, 2008: 168–169). Indonesia has not signed the ILO Convention 169 but the convention has become increasingly influential since many donor agencies (such as DANIDA) stipulate free and prior informed consent (FPIC) (ibid. 169).

In this dynamic context, indigenous groups started to organize and form asso-ciations. Customary communities that had been displaced or dispossessed initi-ated resistance activities, identifying themselves as indigenous people (Peluso et al., 2008: 386). In 1999, approximately 500 indigenous people from different parts of the archipelago met in Jakarta and founded the nationwide indigenous rights organization AMAN. During their initial congress in the same year, they postulated the slogan: "We will not recognize the Nation, if the Nation does not recognize us" (Li, 2001: 645). Today, the organization is well integrated in trans-national networks and supported by international donors (Hauser-Schäublin, 2013: 10). The organization claims to represent all indigenous communities of the archipelago (Bedner and Van Huis, 2008: 167).

REDD+ further strengthens the influence of the organization. AMAN is present at UNFCCC climate change conferences, is regularly involved in side events, and their representatives are regular guests at high-level panel discussions that take place at the margins of UNFCCC conferences. AMAN has success-fully utilized the current focus on forests and the dominant public discourse that frames indigenous groups as forest stewards (Sammukri, 2013: 121–123). With the support of the Norwegian government and the REDD+ Taskforce, AMAN has been involved in a nationwide mapping of indigenous territories and participated in the transnational indigenous rights campaign "No Rights No REDD" (c.f. Hein and Garrelts, 2014: 345).[1]

In 2013, the organization achieved a major success. To further recognition of indigenous rights to land within the state forest territory, AMAN and part-ner organizations forced a decision by the Indonesian Constitutional Court *Keputusan* MK 35/PUU-X-2012) on whether the MoF has the legitimacy to control *adat* forest (Makamah Konstitusi Republik Indonesia, 2012). The court strengthened the rights of *adat* communities vis-à-vis the MoF sig-nificantly, declaring that *adat* forest is no longer under the authority of the MoF (Rachman, 2013). The court ruling acknowledges *adat* communities "as right bearing subjects" (Rachman and Siscawati, 2017: 224), in other words as actors that have "the right to have rights", including rights to land and natural resources (ibid.). This decision is far reaching and has the potential to end the criminalization of customary communities, but it may also increase the probability of ethnic conflicts. At the time of writing, the decision has not been fully implemented.

AMAN successfully shifted political struggle on the recognition of indigenous territories from the local scale to the national scale and to the scale of global climate policy – the UNFCCC climate change conferences. But despite AMAN's claim to represent customary communities across the archipelago and its commitment to self-identification, the organization uses criteria and a verification mechanism to govern its acceptance of members. An AMAN activist responded to a question about whether any community that identifies itself as indigenous could be a member as follows: "Yes, but they have to get a recommendation from other communities which know them and we [AMAN] have our own criteria and verification mechanism, before we accept new members".[2] For instance, AMAN does not fully recognize the territorial claims of the Batin Sembilan in the landscapes of the Harapan Rainforest.[3]

Indigeneity is not a self-evident resource in Indonesia (Rachman and Siscawati, 2017: 237). Indigenous groups have to engage actively in indigeneity politics, and in many cases they need to establish links to NGOs, peasant organizations or AMAN in order to benefit from the indigenous rights discourse (ibid.). Not all groups and not all members of indigenous groups are able to do so.

The reemergence of agrarian movements

Contemporary agrarian organizations in Indonesia have their historical roots in various left-wing organizations formed after independence. The Indonesian Peasant Front (Barisan Tani Indonesia, BTI), for instance, was formed in 1945 and was involved in land occupations on Java, Bali and Sumatra after independence (Peluso et al., 2008: 381; Tuong, 2009: 181). During the New Order era, any resistance against national land allocation policies was suppressed by police and military force. About a 100,000 BTI activists and members of Indonesia's Communist Party (Partai Komunis Indonesia) were killed in the massacres of the 1960s. Consequently, any open peasant resistance came to an end (Peluso et al., 2008: 382).

The first new underground movements emerged in the early 1990s (Peluso et al., 2008: 387). Some cooperated with the environmental justice movement WALHI (Wahana Lingkungan Hidup Indonesia) and with the Indonesian Legal Aid Foundation. Both organizations had been tolerated by the Suharto Regime and were able to start initial multi-scalar campaigns by supporting local farmers and criticizing the lack of implementation of the Basic Agrarian Law (ibid. 384–387). Directly after Suharto's fall, landless people and peasants occupied state forest territory and plantation estates (Peluso et al., 2008: 388; Tuong, 2009: 182). The first explicit peasant and agrarian reform organizations reemerged in the late 1990s. Probably the most influential are Serikat Petani Indonesia (SPI, Indonesian Peasant Union) and Konsortium Pembaruan Agraria (KPA, Consortium for Agrarian Reform). Both organizations share the same political views but KPA is rather urban-based and SPI has a stronger rural basis (Tuong, 2009: 183). Other relevant peasant organizations are STN and AGRA. STN was founded in 1993 in Yogyakarta. The organization aims to support the economic, political and socio-cultural development of peasants, to increase knowledge about "agrarian problems triggered by imperialism and neoliberalism" and to lobby for an agrarian reform that provides access to land, capital and technology for peasants (STN, 2018). AGRA is a rather young agrarian organization established in 2004. The organization considers itself a union for peasants, for ethnic minorities and *adat* communities struggling for agrarian reform (AGRA, 2015). AGRA is a member of the Asian Peasant Coalition and of the International League of Peoples' Struggles (ILPS). All three rural-based peasant organizations (SPI, STN and AGRA) campaign for agrarian reform and all three seem to be relatively close in ideological terms. In an interview, an AGRA activist explained that from his perspective AGRA and SPI both fight for land reform, adding that for AGRA reform is just one of many targets.

SPI was formed in North Sumatra by peasant activists who started to organize in the late 1990s. Since 2003, SPI has been a member of the global peasant organization La Via Campesina. In early years, SPI was mainly involved in advocacy work for peasant farmers, landless peasants and peasants negatively affected by the activities of agribusinesses and state agencies (Purwanto, 2013: 1). From the very beginning, the organization used the Basic Agrarian Law as the legal basis for its campaigns (ibid. 2). SPI considers the law "[. . .] as a progressive law aiming to redress the unfair distribution of agrarian and economic resources [. . .]" (Purwanto, 2013: 2).

SPI considers itself as a community-based anti-capitalistic organization working for socially inclusive land, natural resource and trade policy (Tuong, 2009: 182). At the global level the organization – together with La Via Campesina – campaigns for fair north–south relations, food sovereignty and climate justice (Hein and Faust, 2014: 23; Purwanto, 2013: 2; Tuong, 2009: 185). To reach its goals, SPI organizes mass protests against state agencies and companies and conducts land occupations (Purwanto, 2013: 2). SPI not only campaigns for agrarian reform, by allocating occupied land to landless farmers they also to some extent conduct their own informal agrarian reform. At the global scale SPI is represented by La Via Campesina in most cases. Both organizations organize protests against forest carbon offsetting and REDD+ in the context of international climate change conferences (Hein and Faust, 2014: 25).

In Jambi, the organization has currently 15,000–20,000 members and operates in eight districts.[4] In the Harapan Rainforest, SPI members are involved in land occupations. SPI also supports farmers in a land conflict in the Kerinci National Park, where 3200 farmers are involved in forest conversion and land occupation within the park boundaries (Serikat Petani Indonesia, 2012; Tambunan, 2015).

Concluding remarks on peasant and indigenous rights movements

AMAN and SPI have been quite successful in challenging the state and its apparatuses. They benefitted from the regime change at the end of the 1990s – indeed their informal predecessors probably contributed to the regime change, but that is beyond the scope of this book. Political turmoil and regime change reduced the power of the central state over sub-national public authorities, peasants and indigenous groups. AMAN, SPI and other peasant and indigenous rights organizations were able to develop counter-narratives which were able to challenge the development narratives of the New Order era. In Lukes' and Gaventa's terms, they were able to challenge invisible power. Furthermore, they were able to mobilize communities across the archipelago. In 2012, 1992 indigenous communities were members of AMAN (Alliansi Masyarakat Adat Nusantara (AMAN), 2015). SPI is now present in 14 Indonesian provinces.[5] Hence, both organizations have developed remarkable organizational strength and are able to challenge the visible power of state apparatuses such as the MoEF. Furthermore, they have proved their ability to jump scales and to expand political struggle from the individual household scale of their members to the national (e.g. constitutional court) and the global scale (e.g. the no rights no REDD+ campaign) (Alliansi Masyarakat Adat Nusantara (AMAN), 2015; Purwanto, 2013).

Agro-industrial expansion, land concentration and violence at Jambi's oil palm frontier

Oil palm companies control vast areas of the villages of Seponjen, Kampung Laut, Bungku and Tanjung Lebar (Table 5.1). Peasants are increasingly stuck between large agro-industrial estates and green enclosures. In many cases, the local population claim parts of the corporate concessions, and complain about unfair compensation and unjust benefit-sharing arrangements. Moreover, the

Table 5.1 Land allocated to companies, total village territory, and land used by farmers in the study villages in 2014

Village	"Official" village territory (ha)	Land allocated to companies (ha)	Land used by farmers (ha)
Air Hitam Laut	No data	No companies	No data
Seponjen	16,000	2000[1]	1600[2]
Kampung Laut	12,000[3]	300[4]	825.4[5]
Bungku	~150,000[6]	105,205[7]	No data
Tanjung Lebar	6500[8]	3000	1500[9]

(Source: compiled by the author)

1 Based on interview with key informants in Seponjen, 09.09.2013, 11.09.2013 and 13.09.2013.
2 Based on village government documents, Seponjen (2013: 6).
3 Based on information provided by village government.
4 Based on an interview with a key informant in Kampung Laut, 02.09.2013.
5 Based on information provided by village government, Kampung Laut (2012).
6 Based on Zainuddin.
7 Ibid. and Mardiana (2014: 16).
8 Based on Polsek Sungai Bahar (2011).
9 Based on Polsek Sungai Bahar (2011).

land conflicts over access to and control of the Harapan Rainforest are deeply entangled in conflicts with the various oil palm companies surrounding the conservation concession and local communities. The conflicts involve the same actor groups, and the oil palm company PT Asiatic Persada has pushed the Batin Sembilan and other smallholders into the remaining forest patches of the Harapan Rainforest (Beckert et al., 2014: 86; Colchester et al., 2011: 15; Hein et al., 2016). The conflict with PT Asiatic Persada is one of the longest ongoing land disputes in Jambi, involving one of the most controversial plantation developments in the province. The conflict has led to a number of injuries, to at least one casualty, and to the destruction of a number of the farmsteads, old graveyards and forest gardens of the indigenous Batin Sembilan.

The previous concession owner PT Bangun Desa Utama (PT BDU) received a 20,000 ha concession in 1984 that overlapped with land claimed by the village communities of Bungku, Tanjung Lebar, Markanding and Pompa Air (the latter two were not included in the villages visited for this study). It is very likely that the Batin Sembilan and the populations of the affected hamlets had no say regarding the issuance of the plantation concession (*Hak Guna Usaha*); according to a key informant in Bungku, they were not consulted prior to the plantation's implementation.[6] The same informant stated that the company destroyed rubber plots and fruit gardens: "[. . .] originally this was all community land. We had rubber, durian, and *jernang*[7] (dragon blood). It was land of the people but the company destroyed it with bulldozers". A Batin Sembilan elder explained that the plantation estate was constructed with support from the military and the police: resistance was impossible.[8] A significant part of the Batin Sembilan population now living in the Harapan Rainforest once lived in areas now used by PT Asiatic Persada.

Although the conflict with PT Asiatic Persada has not yet been resolved, rescaling processes and scale jumping to the transnational scale facilitated by the NGO SETARA has strengthened the position of local communities vis-à-vis the company. SETARA's transnational partners, such as the NGOs Robin Wood and Rettet den Regenwald, organized protests in Germany against the company and its customer Unilever (Beckert and Keck, 2015; Robin Wood, 2011, 2014). PT Asiatic Persada's former owner, the Singapore-based agrobusiness group, Wilmar International, recently sold the company to avoid losing its certification from the Roundtable for Sustainable Palm Oil (RSPO)[9] (c.f. Beckert et al., 2014; Steinebach, 2013b). At the time of my field research, members of the local activists group SAD 113 had occupied and managed approximately 241 ha of the plantation. The activists claim that they descend from three pre-colonial hamlets (Tanah Menang, Pinang Tinggi and Padang Salak).

In more recent smaller corporate plantation developments in the villages of Tanjung Lebar, Seponjen and Kampung Laut, local public authorities were directly involved in the permit process – although conflicts and disputes also occurred. In Tanjung Lebar, for instance, PT Bahar Pasifik received a permit from the *Lembaga Adat* of Tanjung Lebar, which managed customary land that the Batin Sembilan community had received from the provincial government in the early 2000s.[10] The *Lembaga Adat's* permit required the company to set aside land for a smallholder scheme but once the company received the HGU from the district government, it refused to allocate land to smallholders.[11] Some of the displaced smallholders then moved into the Harapan Rainforest. In this case, the company applied scale jumping to the district scale of regulation to avoid the costs of running a smallholder scheme. Key informants have reported a similar case in Seponjen at the margins of the Berbak Carbon Initiative.

Peasant resistance against the companies occurred in both hidden and in open ways. In the case of PT Asiatic Persada, peasants openly occupy parts of the concession and have marked the occupied areas with flags and boundary stones. In other conflicts, peasants rather relied on hidden resistance strategies, such as sabotage. In Seponjen, for instance, peasants stole the keys of land machinery to hinder land conversion for a new plantation estate.

Corporate oil palm plantations are powerful actors in the landscapes adjacent to the Harapan Rainforest and the Berbak Carbon Initiative. They hold large oil palm concessions which in many cases overlap with land claimed as customary land by peasants and indigenous groups (Beckert, 2017; Beckert et al., 2014, Beckert and Keck, 2015; Steinebach, 2013b;). Many villagers were displaced in the context of the plantation developments while others lost their land and then started to work for the companies. The village of Bungku, which inhabitants call the Village of 1001 Problems has been in a situation of permanent conflict and emergency for more than 20 years. Peasant and indigenous rights to land and livelihood strategies are permanently at risk. In this conflictive political context, peasants and indigenous groups engaged in collective learning about how to deal with the permanent threat of losing access to their land, formed local resistance movements and established links with political parties and local and transnational NGOs. The established networks and social capital again became relevant in the context of the implementation of the Berbak Carbon Initiative and the Harapan Rainforest. The new conservation initiatives challenge the ongoing agricultural expansion driven by peasant migrants allied with indigenous leaders and village governments. Peasants met in Bungku, for example, complain that they first lost their land to oil palm companies, and now their livelihood is threatened by attempts to protect the "lungs of the earth".

Conservation vs. agrarian reform: conflict between SPI and the Harapan Rainforest

The first of the four transnationalized agrarian conflicts about access to and control of the "lungs of the earth", described in this chapter, is probably the most intense and violent conservation conflict in Jambi. The conflict involves members of Serikat Petani Indonesia (SPI) allied with the transnational partner La Via Campesina, the conservation company PT REKI and various state apparatuses (the Ministry of Forestry, police and military). Despite PT REKI's presence, peasants affiliated to SPI have been able to expand forest conversion and settlement formation (Figure 3.1). Peasant farmers in Sungai Jerad occupy and use land for agricultural purposes that the MoF has allocated to the conservation company PT REKI. Accordingly, SPI members openly challenge the hegemony of the MoF. SPI considers the occupation of state forest as a legitimate response to colonial and post-colonial policies of dispossession (Hein et al., 2016). SPI members lobby for the release of the claimed area from PT REKI's conservation concession and for a reclassification of the area to non-forest land.

Peasant resistance in the context of the Harapan Rainforest project is a spatial and territorial practice that challenges the scales of meaning and regulation constructed by PT REKI and by the MoF (Moore, 1998; Turner and Caouette, 2009; Towers, 2000). Territorial, in the sense that SPI and PT REKI have established specific regulations and border demarcations for the spatial units they control. The conservation company has formulated conservation regulations (explained in Chapter 4) and SPI the three T rules (explained in Chapter 3). The conservation territory and territory controlled by SPI are marked with signs, turnpikes and flags that represent power relations inscribed in the landscape.

Peasant resistance facilitated by SPI not only challenges implementation of the Harapan Rainforest project, but SPI's resistance can also be considered as being part of a larger and multi-scalar resistance campaign. At the village scale, SPI seeks to tackle the local impact of national land allocation policies through land occupations. At the national scale, SPI seeks to change the national legal framework by organizing protest in urban centers. At the global scale, SPI seeks to challenge the implementation of transnational forest carbon offsets (c.f. Hein and Faust, 2014; Hein et al., 2016).

Conflict history

According to SPI members, the conflict started in 2008 and intensified in 2010. SPI members reported that in 2010 the forest police and staff of PT REKI started patrolling in the settlements and announced that the land is now part of the concession of the conservation company.[12] SPI leaders argued that the conservation company has not conducted FPIC and has denunciated SPI members as encroachers and illegal loggers.[13] In contrast, PT REKI argues that SPI members were not willing to participate in any consultations.[14] The conflict escalated in 2012. Both parties accuse each other of kidnappings and the destruction of property. In particular, SPI members accuse the forest police, BRIMOB, the military and private informal security forces of having destroyed farmsteads (Lang, 2012a; Usman, 2012; Wirasapeotra and Octavian, 2012).[15] Furthermore, SPI members complain that the raids of BRIMOB, the army and forest police only targeted smallholders and not the larger plantation estates held by elites and politicians within the concession.[16] Since 2013, the conflict has calmed somewhat but is not yet resolved, and both parties accuse each other of not being willing to participate in conflict mediation. According to SPI activists and the staff of Burung Indonesia who I met in August 2016, the situation has not changed. However, PT REKI argues that SPI has lost control over its territory and that increasingly it is land speculators who are benefitting from the situation (Suprapto, 2016).

The extent to which specific actors are involved in organized forest crimes and in human rights violations is hard to assess and goes beyond the scope of this book. Nevertheless, PT REKI accuses SPI members in particular of being paid by logging companies to cut valuable timber in the conservation concession.[17] SPI accuses PT REKI of paying informal security staff who have been involved in human rights violations.[18] According to a local newspaper, the Indonesian National Commission on Human Rights (Komisi Nasional Hak Asasi Manusia, KOMNAS HAM) has found indications of the occurrence of human rights violations in the context of the conflict (Ferdiyal, 2013).

Actor mapping

The main actors in the conflict apart from PT REKI and SPI (Figure 5.2) are the MoEF, sub-national forest agencies, various security apparatuses (e.g. the police and the Indonesian Armed Forces/*Tentara Nasional Indonesia* (TNI)),

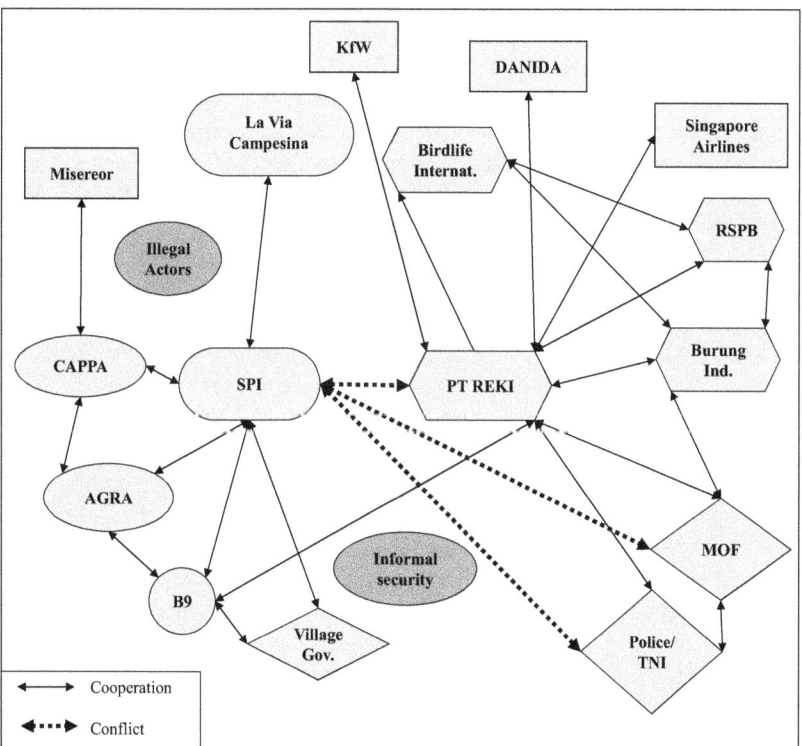

Rounded rectangle: SPI and partners, oval: other NGOs, diamond: state actors, hexagon: PT REKI and partners, rectangle: donors, circle: Batin Sembilan elites, dark oval: illegal actors and informal security staff (Source: own draft, based on interviews with involved stakeholders).

Figure 5.2 Actor mapping of the conflict between SPI and PT REKI on access and control of the Harapan Rainforest concession.

(Source: compiled by the author)

La Via Campesina as SPI's transnational partner, the transnational shareholders of PT REKI (Burung Indonesia, RSPB and Birdlife International) and PT REKI's foreign donors (e.g. DANIDA and KfW).

Actors such as Singapore Airlines, DANIDA and KfW are of course not directly involved in the local confrontations. Nevertheless, KfW as the implementing organization of the German International Climate Initiative (IKI) has not only funded the project, but has also articulated verbal and written support for PT REKI in public debates on the conflict (c.f. Lang, 2012a).[19] Especially on the online platform REDD-Monitor, a debate between SPI and supporters of PT REKI was ongoing between 2008 and 2012 (Lang, 2015). KfW has participated in these debates, supporting PT REKI. In a joint response formulated by KfW, IKI and the Federal Ministry for the Environment, Nature Conservation and Nuclear Safety (BMUB), the German donors consider the activities of SPI members "[. . .] illegal land-grabs" (Lang, 2012a).

It is also important to consider that, as mentioned in Chapter 4, IKI neither had strict social safeguards in place nor stipulated FPIC. DANIDA, in contrast, explicitly states that "[. . .] when designing and implementing environment and climate initiatives, the principles of free, prior and informed consent must be adhered to" (DANIDA, 2004). AGRA, however, complains that PT REKI had not conducted FPIC in the area of Pangkalan Ranjau prior to DANIDA's recent grant (AGRA, 2017). In 2015, PT REKI announced its new human rights-based approach, including its commitment to FPIC. However, the extent to which FPIC could be implemented adequately when the project had already been running for more than five years remains questionable (REKI, 2015). The implementation of FPIC and social safeguards would have probably helped to mediate the ongoing conflict in the first place.

DANIDA did not participate in the public debates on REDD-Monitor. However, DANIDA, as the German donors, actively reproduced the social construction of peasant migrants as illegal actors by using the term "encroachers" in annual reports of its Environmental Support Programme (DANIDA, 2015, 2016). In 2015, DANIDA wrote: "However, serious threats to the success of Hutan Harapan are still present. Most importantly, the continued encroachment by in-migrants in the northern and eastern part of the concession" (DANIDA, 2015: 8). Directly referring to the area occupied by SPI, DANIDA wrote in 2016 that the lack of law enforcement "[. . .] could endanger further Danish support" (DANIDA, 2016: 25). In addition, a number of statements found in the report indicate that the Danish agency supports a rather militarized conservation approach. The 2016 report highlights that new forest posts and GPS-based monitoring have been established and that twice a month PT REKI conducts joint patrols involving the police, army and district forest agency (DANIDA, 2016: 23).

Singapore Airlines has contributed to a trust fund of the Harapan Rainforest project and labels itself as the "[. . .] exclusive airline partner for the Harapan Rainforest Initiative [. . .]" (Singapore Airlines, 2015). The carrier does not receive carbon credits in return but considers its support as being part of a commitment to improve environmental performance (ibid.). Following this argument, Singapore Airlines' contribution to the trust fund can be considered as being at least an indirect and unquantified environmental offset. An expert from KfW described the financial contributions to the trust fund as "virtual certificates" which do not lead to the production of regular carbon credits.[20] In the carrier's sustainability report, the Harapan Rainforest is presented as a success case, thus concealing the complexity of the ongoing local conflicts.

Even though not directly involved in project implementation and management, the support of these main donors goes beyond just financial contributions. Especially KfW and DANIDA reproduce existing power asymmetries among the conservation company, the MoEF and peasants by further delegitimizing the presence of peasant migrants in the state forest territory through use of the term "encroachers" and by calling for stricter law enforcement to "solve" the problems related to in-migration (DANIDA, 2015, 2016). Moreover, in one way or another, their involvement constitutes the transnational and multi-scalar dimension of this conflict. These multi-scalar and transnational aspects of the conflict have a dialectical character, they provide additional agency for transnational protest but at the same time additional resources for stricter law enforcement and militarized conservation approaches. The additional agency was used by SPI and facilitated by La Via Campesina. SPI jumped to the UN scale to raise concerns about REDD+ and the Harapan Rainforest project at the global scale, in this specific case at the conference of the parties at the UNFCCC in Poznan in 2008 (La Via Campesina, 2008).

After the initial clashes in 2012 and 2013, state actors seemed to have reduced their presence in the area controlled by SPI. In a recent DANIDA report, the donor complained that the MoEF does not have the financial means to conduct law enforcement in the area occupied by SPI (DANIDA, 2016: 25). During a workshop in Jambi that I organized with partners of the Agricultural University Bogor and the University of Jambi in August 2016, actors involved in the conflict also argued that the weak presence of the state is a major cause of PT REKI's "problems". Others highlighted that the local communities require legalization by the national government.

Other relevant actors in the conflict arena are the NGOs Yayasan CAPPA and AGRA and PT REKI's transnational shareholders Burung Indonesia, RSPB and Birdlife International. The NGO Yayasan CAPPA supports migrants and Batin Sembilan in land conflicts in Bungku and regularly coordinates campaigns with SPI. AGRA also closely collaborates with CAPPA and supports Batin Sembilan groups. In September 2013, the peasant organization AGRA started to map the territorial claims of Batin Sembilan in the area of Pangkalan Ranjau within the Harapan Rainforest. AGRA highlighted that the Batin Sembilan had become a minority in the Sungai Jerad area and that they needed support in the negotiations with PT REKI and SPI.[21] The Batin Sembilan groups living close to the SPI settlements (B9 in Figure 5.2) are not directly involved in the conflict. However, as described in Chapter 3, Batin Sembilan elites granted access to land to SPI members in the first place. Though, many non-elites were not able to benefit from the land transactions and some complained about the rapid expansion of SPI settlements. Recently, observers reported tensions between the Batin Sembilan group supported by AGRA in the hamlet of Pangkalan Ranjau and SPI. The leader of the Batin Sembilan group in the hamlet argues that SPI is occupying land within the customary land of his group without having permission to do so (Suprapto, 2016). PT REKI's shareholders have slightly different positions in the conflict. From the very beginning the RSPB was in favor of a strict fortress conservation approach and aimed to relocate or evict all settlers living in the concession.[22] Burung Indonesia, in contrast, was more in favor of a mediation-based solution to the conflict (Silalahi and Erwin, 2013).[23]

A multi-scalar conflict over meanings, access to and control of the Harapan Rainforest

Although the main cause of the conflict is the overlapping land rights legitimized by different public authorities and competing interests in mutually exclusive land-use practices (e.g. farming vs. conservation), the transnational linkages and the strategies employed by the various conflict parties make the case even more complex. Both parties seek to situate their activities within broader scales of meaning and refer to competing scales of regulation. SPI attempts to construct explicit linkages between the confrontations in the Harapan Rainforest and national agrarian and global climate justice debates. SPI stands in the tradition of the socialist zeitgeist of the President Soekarno era, and is struggling for the reconstruction of national scales of meaning and regulation based on the key role that land plays in achieving welfare and social justice in rural areas (Hein and Faust, 2014). Interviewed SPI members settling in the Harapan Rainforest argued that the Basic Agrarian Law (BAL) and the constitution provides legitimacy for their land claims: "[. . .] the BAL from 1960, no. 5 clarifies the right of the people to land [. . .] we did not have a formal permit for the land here but we have a strong law on our side".[24]

The settlements within the Harapan Rainforest provide land for SPI members, hence they strengthen the attractiveness of the SPI and the potential for political mobilization for national campaigns (Hein et al., 2016). At the transnational scale, SPI and its ally La Via Campesina use the settlements and the associated land conflicts with PT REKI to underpin transnational campaigns against carbon offsetting and REDD+ (Hein and Faust, 2014; Hein et al., 2016; La Via Campesina, 2012). SPI rejects the commodification of nature in the name of the "green economy", conservation and REDD+. Furthermore, SPI – in line with La Via Campesina – argues that developed countries should reduce emissions domestically and not through the enclosure of large forest areas in the Global South.[25] The head of SPI Jambi stated in simple terms, "Traditional agriculture creates a healthy environment but instead displacements of farmers and [. . .] detention of farmers occur, thus REDD is equivalent to the displacement of farmers".[26] SPI members settling in the Harapan Rainforest have argued, "PT REKI takes care of the lungs of the earth by re-greening, but who takes care of the lungs of the people?"[27] And, "I have just one question the carbon goes overseas, what is more important the carbon or the community living here? [. . .] When we have carbon we do not have people here, who is the carbon for then?".[28]

PT REKI and partners especially stress the global and local relevance of the Harapan Rainforest project. Birdlife International (2008: 4), as one of PT REKI shareholders, stress in a project report that "[. . .] the lowland rainforests of Sumatra are of particular importance and rival the Amazon in terms of species" (ibid). Furthermore, the report highlights the "carbon value" of the project (ibid. 6). A former lobbyist of Burung Indonesia highlighted the singularity of the ecosystem and the presence of large mammals and asked, "How can we commodify conservation to increase protection?".[29] For DANIDA the greenhouse-gas reduction component seems to be of specific importance. A specific goal of DANIDA's environmental support program is to reach a carbon storage target of 130–150 tons per ha in the Harapan Rainforest (DANIDA, 2015: 41). Officially, PT REKI has disassociated itself from REDD+ and has not pursued initial ideas to sell credits on the voluntary carbon markets. This decision was probably taken to avoid controversies on benefit sharing and to wear down SPI's anti-REDD+ campaigns. The management of PT REKI (as described in Chapter 4) had no consistent position on financing the project directly via carbon markets. In 2013, a project manager stated that currently the company has other problems than engaging in carbon trade.

In the direct stand-off with SPI the conservation company PT REKI employs rather a legalistic approach, highlighting that SPI's activities and presence violates the Forest Law. A staff member of PT REKI for instance argued, "[. . .] SPI are bad guys, they are involved in illegal logging and land trade, they are criminals", adding that, "fortress conservation is the only way to conserve this forest".[30] Moreover, PT REKI stressed that the ecosystem restoration concession allocated by the MoF is the only property right in place issued by a national-scale authority. Thus, PT REKI challenges property rights issued by other public authorities (e.g. village-scale land titles) and reproduces the state forest territory and its concession system as scales of regulation.

In spite of all their differences, both parties strategically employ indigeneity politics and both parties argue that they contribute to forest conservation. PT REKI and partners argue that the project is locally relevant especially for the Batin Sembilan. The project area can be considered as a last resort for the Batin Sembilan providing "[. . .] them with the option of continuing to reside in a forest environment" (Birdlife International, 2008: 5). Singapore Airlines argues, "Harapan also plays a significant role in terms of engaging the local indigenous community. Around 800 people from the Batin Sembilan indigenous group depend on the forest for their livelihood" (Singapore Airlines, 2010). The German NABU (Naturschutzbund Deutschland), a member of Birdlife International, argues that the survival of the Batin Sembilan depends on the protection of the Harapan Rainforest (NABU, 2010). In contrast, a field officer of PT REKI considers migrants "as a major challenge for conservation". PT REKI considers the land claims of the Batin Sembilan as more legitimate than those of migrants.[31] In addition, PT REKI uses the presence of the Batin Sembilan and their alleged role as "forest-dependent people" (Birdlife International, 2008: 5) to legitimize forest conservation. SPI, as described in Chapter 3, claims to have received a permit from the hamlet head of Mangkubangan/Pangkalan Ranjau and from Batin Sembilan leaders. SPI argues, as the conservation company does, that their activities contribute to conservation. In this sense, the three "T" rule and the oil palm ban can be considered as being part of SPI's strategy to increase the legitimacy of their activities. SPI members have argued that the forests of Sungai Jerad and Bukit Sinyal were destroyed by the logging company PT Asialog. Following this argument, rubber agroforestry systems planted by the SPI settlers are contributing to forest rehabilitation. Furthermore, the head of an SPI basis argued that the area was *lahan tidur* (which is a term describing degraded unproductive land) before SPI came to the area. SPI brings the land back to productivity, he suggested.[32] The conservation company rejects these arguments and refers to Landsat and Rapid Eye images that indicate ongoing forest conversion and oil palm cultivation (Lang, 2012b).[33]

The conflict about Kunangan Jaya I: defending village expansion

The conflict about the hamlet of Kunangan Jaya I is again caused by overlapping property rights legitimized by different authorities. Even more importantly, the conflicts concerning Kunangan Jaya illustrate power shifts and the new fragile scalar structure of the post-Suharto period. The emergence of a village scale of land tenure regulation and of attempts by Batin Sembilan groups to reestablish their customary territories after Suharto's fall has been challenged by the MoF, who allocated the Meranti-River-Kapuas-River forest block to the conservation company PT REKI, thus neglecting the presence of local communities and pre-existing access and property relations (cf. Hein et al., 2016, and Chapter 3).

The hamlet of Kunangan Jaya I has more than 1200 inhabitants and consists of many dispersed settlements mainly located within the concessions of PT REKI and PT Asiatic Persada (Wirasapeotra and Octavian, 2012: 7). Each settlement has its distinct history. I will only focus on conflicts involving the TSM settler community (introduced in Chapter 3) and different Batin Sembilan groups settling in the northernmost part of the Harapan Rainforest concession. The conflict over Kunangan Jaya I has multiple dimensions and multiple strands. Depending on the perspective assumed, the conflict can be framed as a conflict between different apparatuses of the state (e.g. involving village governments, MoEF and sectoral agencies), as a transnational conflict over indigenous rights, and as a conflict between local communities interested in maintaining access to agricultural land and actors interested in conservation and climate change mitigation. Moreover, the different local communities involved in the project are very heterogeneous. Different community members and community leaders have different abilities, material resources and interests, and thus follow different strategies in the conflict with PT REKI. The TSM community and some of the Batin Sembilan communities seek the release of their settlement out of the state forest (*enklave*). Others have accepted conditional land tenure agreements with PT REKI.

Conflict history

The first land conflicts over access to and control of the landscapes of Kunangan Jaya I date back to the period when the area was managed by the logging company PT Asialog. In the 1970s, in the course of the implementation of the logging concessions, the first Batin Sembilan families of Kunangan Jaya who practiced shifting cultivation were forced to leave the concession. The more recent conflicts related to the designation of the Meranti-River-Kapuas-River forest block for ecosystem restoration and to the allocation of the forest block to PT REKI started in 2007. In 2007, the District Forest Agency and the forest police started a first attempt to relocate the TSM community. After initial negotiations, the district forest agency tentatively accepted their presence and informed settlers that they should plant rubber instead of oil palms.[34] A key informant reported that the conservation company started to intimidate peasants in the area even before the company received its concession.

In 2010, after the conservation company officially received its concession, the company supported by the heavily armed mobile police brigade (BRIMOB) and the forest police entered the community, urging people to leave the concession and to abandon their farmsteads and plantations within two months.[35] In other parts of Kunangan Jaya I (e.g. the area around Sungai Kandang and Simpang Macan), key informants complained that forest rangers of PT REKI destroyed their oil palm plantings and farmsteads.[36] A few settlers in Kunangan Jaya I reported that PT REKI offered compensation to those peasants that leave the area "voluntarily". Some accepted the compensation and left the area; others accepted the payments, left and moved back later. In almost all the settlements, community members complained that PT REKI failed to conduct FPIC. The members of the different communities in Kunangan Jaya I reported that they were only informed in the course of the police interventions that agriculture and settlement is prohibited within the state forest.[37] After the police intervention, a farming group from Kunangan Jaya I (PERTAMA, Persatuan Tani Mandiri) supported by the NGO Yayasan CAPPA organized demonstrations in front of the Governor's Palace and in front of the provincial parliament in Jambi city. Only after the demonstrations in Jambi city did PT REKI and the forest authorities conduct the first community consultations (*socialisasi*) in the TSM settlement.[38] After 2013, the conflict calmed down, but various negotiations and conflict mediation aiming to find a solution among the different communities, PT REKI and forest agency have continued until the present day.

Actor mapping

An important feature of the conflict over Kunangan Jaya I is the fragmented character of the community (Figure 5.3). The community of Kunangan Jaya I consists of different factions following different interests and strategies. Conflicts occur not only between community members and the conservation company PT REKI and state actors. Conflicts also occur between different factions of the community of Kunangan Jaya I. The TSM settlement, for instance, consists of at least two opposing factions: members of the farming group PERTAMA and peasants associated with Pak Kumis. Members of the farming group PERTAMA and a number of peasants led by a neighborhood head (Kepala RT) felt betrayed by Pak Kumis and the other organizers of the settlement. PERTAMA members argued that they were not aware that the TSM settlement is located within the state forest, thus within the Harapan Rainforest. Some settlers accused Pak Kumis of being involved in illegal logging and reported that he made high profits from land trade instead of building functional infrastructure

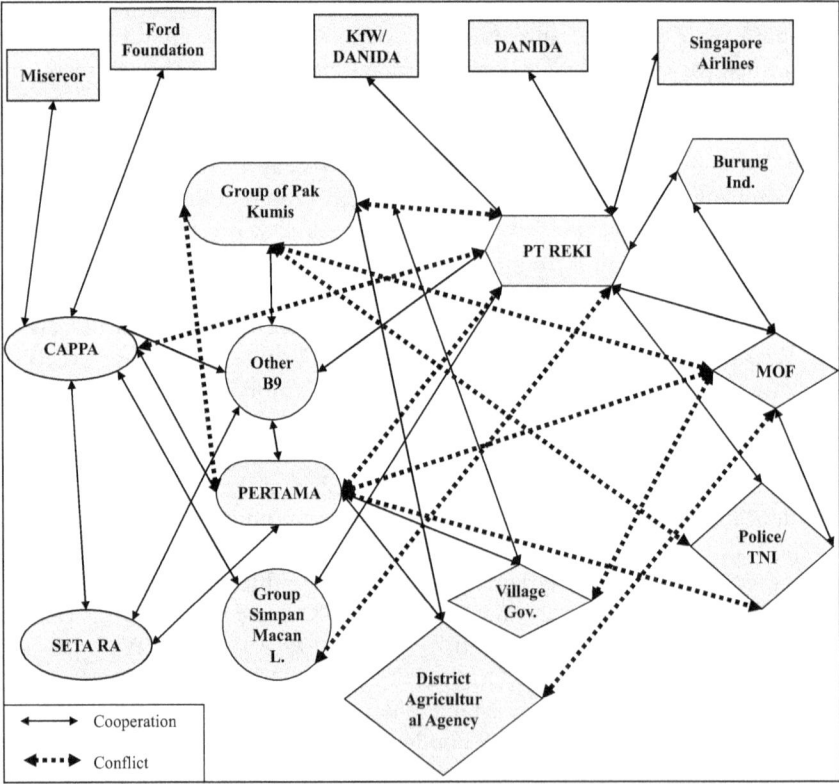

Rounded rectangle: peasant groups, oval: NGOs supporting local communities, diamond: state actors, hexagon: PT REKI and partners, rectangle: donors, circle: Batin Sembilan groups, (Source: own draft, based on interviews with involved stakeholders)

Figure 5.3 Actor mapping of the conflict about Kunangan Jaya I

and houses for Batin Sembilan families.[39] A Batin Sembilan elder complained that his family lost large parts of their customary land (*Wilayah Adat*) to Javanese, Sundanese and Minangkabau migrants because of the formation of the TSM settlement. What is left over, he argues, is too small to provide a safe livelihood for future generations.[40]

PERTAMA receives support from the NGOs Yayasan CAPPA and Yayasan SETARA. PERTAMA members also participate in the conflict mediation meetings with PT REKI. Their most important concern is to get their land released out of the conservation concession. A member of PERTAMA pointed out: "[. . .] we reject the community development program (*program kemitraan*), [. . .] we just want to maintain our land".[41]

Pak Kumis and peasants associated with him have no clear position in the negotiations with the conservation company. When interviewed in 2012, Pak Kumis argued that the conflict has to be solved by the government according to the existing legal framework. He argues that mediation is not necessary since the issuance of the conservation concession by the MoF was a failure because the ministry ignored the presence of the TSM settlement. His group has rejected the support of NGOs but some members have participated in the mediation process. His group supports PT REKI's community development zone.[42]

A third faction of TSM settlers consists of those households that do not feel represented by Pak Kumis's group or by PERTAMA, or are not participating in the mediation process. A recently arrived Javanese household head, for instance, stated that he has no knowledge about PT REKI and the conflict at all.[43]

The NGO Yayasan CAPPA has an important role in the conflict about Kunangan Jaya I. First, the NGO has supported the TSM community in the mediation process. Second, the NGO has conducted workshops on REDD+ (*socialization*) in Kunangan Jaya I and II. Third, the NGO has acted as an up- and downward translator (Pasgaard, 2015: 113). The organization has translated global policies (e.g. REDD+) to local spaces and has aligned local concerns (e.g. risk of displacements) with transnational debates on the rights of local and indigenous communities (e.g. FPIC) (c.f. Mannell, 2014: 2; Pasgaard, 2015: 113). Through organizing demonstrations, posting letters of complaint from Batin Sembilan on REDD-Monitor, and contacting PT REKI's donors, CAPPA shifted the land conflict about Kunangan Jaya I to the provincial, national and transnational scale. Yayasan CAPPA receives funding from the Ford Foundation to mediate and solve land conflicts, and from Misereor to explain REDD+ to local communities. In an allusion to Yayasan CAPPA's funding sources, a staff member of PT REKI argued that the NGO is creating and exaggerating conflicts in order to access funding.[44]

Yayasan SETARA is the second NGO supporting the community of Kunangan Jaya I. SETARA, as a member of the transnational Roundtable on Sustainable Palm Oil (RSPO), mainly supports Batin Sembilan in the conflict with PT Asiatic Persada. SETARA especially criticizes PT REKI's criminalizing of the settlers as illegal encroachers.[45] SETARA cooperates intensively with CAPPA.

Burung Indonesia, as one of PT REKI's shareholders, has conducted additional meetings with community members in Kunangan Jaya I. The KfW is, of course, not directly involved in the conflict. However, in contrast to the statements KfW made about SPI members, staff of KfW argued that the development bank has advocated the exclusion of the hamlets of Kunangan Jaya I and II from the conservation concession from the very beginning and wants the area to be reclassified as non-forest land.[46]

The conflict about Kunangan Jaya I is not only a transnationalized land conflict between peasants and the conservation company PT REKI. The conflict can also be considered as a conflict between different apparatuses of the state. Different settlement projects within the borders of the Harapan Rainforest (e.g. TSM in Kunangan Jaya I, Camp Gunung in Kunangan Jaya II, and Tanjung Mandiri in the village of Tanjung Lebar) have been actively supported by different state agencies. The formation of the settlements has been supported by village governments and by the agricultural agencies of Batang Hari and Muaro Jambi, and by the education agency of both districts. The allocation of corn and soy seeds to farming groups in Kunangan Jaya I by the Agricultural Agency of Batang Hari provided legitimacy for the settlers, supplying them with in-kind start-up funding, but this violated Forest Law 41/1999. The allocation of village-scale land titles by the village government of Bungku facilitated the transaction of land claimed by the MoEF, but again violates existing de jure regulations for forest land. Conflicts have also occurred within the forestry apparatus. The District Forest Agency of Batang Hari, for instance, tolerated the TSM settlement after negotiations with settlers in 2007. Only a few years later, after PT REKI received the conservation concession, the forest police and the mobile police brigade sought to evict the settlers.

Conflict mediation and collaboration agreements

In 2011, pre-negotiations for conflict mediation started and a participatory land tenure mapping was conducted. The mapping involved community representatives, the District Forest Agency of Batang Hari, the Provincial Forest Agency of Jambi, activists of Yayasan CAPPA and staff of the conservation company PT REKI. Official mediation started in 2012 after PT REKI and representatives from Kunangan Jaya I agreed on 20 rules as a basis for mediation. The agreed rules included, inter alia, the prohibition of land transactions, commercial logging, and land clearance outside the area identified in the participatory mapping conducted by PT REKI, the settlers, and the forest service (Kesepakatan Terhadap Prasyarat Mediasi Antara PT. REKI Dengan Warga RT 11 2012). The agreement permitted settlers to maintain their land but prohibited further agricultural expansion (ibid.). The agreement and the participatory land tenure mapping can be considered as an unofficial conditional land tenure scheme since it provides tenure security if peasants accept the negotiated rules.

At the same time, the Provincial Forest Agency received a request from community members of Kunangan Jaya I and II asking that the area be released out of the conservation concession and out of the state forest. The request was rejected. A staff member argued that the social and biophysical criteria[47] were not met. He stated explicitly that the multi-ethnic character of the settlement and the presence of non-indigenous communities was an important factor for the rejection. "[. . .] Most of the people in the area were not SAD,[48] they are from Medan, they have Batak[49] names, why do they ask for an enclave, they are outsiders, the area is still covered by forest, it's not their land".[50]

In 2012, two options were suggested by the MoF to solve the conflict with the TSM community of Kunangan Jaya I. In the first option, the TSM settlement would remain within PT REKI's conservation concession. The area would be designated as a community development zone (*mitra zone*). Settlers would receive conditional land tenure. The allocation of land rights of up to 5 ha were under discussion. In the second option, the TSM settlement would be released out of the conservation concession but would remain part of the state forest. Settlers would receive smallholder forest concessions of 2 ha per household (HTR, Hutan Tanaman Rakyat). Later on, the MoF itself rejected the second option arguing that HTR schemes within a limited production forest (*Hutan Produksi Terbatas*) would violate Forest Law; hence, they urged PT REKI to implement the first option.[51]

The Batin Sembilan dominated settlements of Simpang Macan Luar and Dalam were not targeted by the interventions of the forest police and mobile police brigade. As outlined in the section on the conflict with SPI, PT REKI considers the land claims of Batin Sembilan communities as more legitimate than those of non-local ethnic groups. A staff member of PT REKI stated, "REKI has never refused Batin Sembilan communities".[52] The conflicts between PT REKI and Batin Sembilan have a much lower intensity than those with non-local ethnic groups. PT REKI has already negotiated conservation agreements with the community in Simpang Macan Dalam. Negotiations with the community of Simpang Macan Luar were still ongoing in 2013 (Buletin Batin Sembilan, 2013: 26). Parts of the Batin Sembilan community of KM 35,[53] located close to PT REKI's main camp, reject negotiations with the conservation company arguing, "[. . .] I do not want to participate in PT REKI's program, frankly speaking I do not participate, I might lose out, and I have customary rights to this land".[54] Peasant migrants and some Batin Sembilan criticized the conditional land tenure and conservation agreements. A Batin Sembilan elder said that he refused to be resettled and also refused the conservation agreements because he has rights based on customary law (*adat*).[55] A settler told me that he did not want to sign any agreements with PT REKI, he just want to keep his land.

In 2013, some community leaders complained that the negotiations with PT REKI were broken off by the company without them being informed. Negotiations resumed in 2015, and in 2016 PT REKI was able to negotiate collaboration agreements. PT REKI has signed agreements with approximately 50 percent of the households of Pak Kumis' group (DANIDA, 2016; PT REKI, 2018). The agreement permits them to use 153 ha of land for rubber cultivation. In addition, an agreement was reached with groups represented by a neighborhood head and with a number of the Batin Sembilan groups of Kunangan Jaya

I. The agreement made with the Batin Sembilan includes 1465 ha of land for 390 persons and permits agroforestry, horticulture and the cultivation of annual food crops (REDDplusid, 2016).

The various agreements have solved most of the land conflicts in the area of Kunangan Jaya I. Through negotiating and signing the agreements, the different peasant and indigenous groups have accepted PT REKI's authority over land allocation and land use in the landscapes of the Harapan Rainforest. The agreements challenge the territorial claims of village governments and indigenous leaders from Bungku as well as the village-scale forest conversion and settlement schemes and associated village-scale land titles. Peasants and indigenous groups were not successful in lobbying for the reclassification of state forest territory. They did not receive land rights legitimized by the National Land Agency or smallholder concessions (HTR) from the MoEF. Nevertheless, some peasant migrants who have not been able to claim land rights based on ancestry and kinship have at least received rights to land legitimized by the conservation company.

Competing scales of regulation, conflicts between state apparatuses and indigeneity politics

The village government of Bungku expanded its competencies formally and spatially by facilitating settlement formation within the state forest and issuing village-scale land titles for the settlers. Migrants, the village government and Batin Sembilan elites constructed alternative scales of meaning based on the reconstruction of the development narratives of the President Suharto era and on the idea of re-establishing pre-colonial ethnic territories. Arriving migrants actively reproduced the authority of the village government and of the village scale of regulation by requesting land and land titles.

PT REKI's conservation concession is located within the state forest territory and was issued by the MoF. At the same time, the concession is located within the ethnic territory of the Batin Sembilan. By requesting the concession, PT REKI contributed to the reproduction of the state forest as a national scale of meaning and regulation controlled by the MoF, one that challenges the alternative scales constructed by other actors (e.g. Batin Sembilan elites). This conflictive scalar arrangement was the starting point of the struggles over Kunangan Jaya. The struggles between the various actors involved in the conflict induced further scalar restructuring, e.g. up-scaling through shifting the conflict to the emerging global scale of forest governance.

Moreover, the conflict highlights the relevance of transnational and multi-scalar resistance in the context of REDD+. The active use of global norms and regulations by different actors to solve land conflicts contributes to the construction of a global scale of regulation for governing the world's forests. In the context of an emerging global scale of forest governance, global norms such as FPIC are becoming increasingly relevant in local land conflicts. The NGO Yayasan CAPPA plays a key role since the NGO has knowledge on relevant international debates, speaks the international development jargon, and is able to communicate directly with donors. Consequently, Yayasan CAPPA is able to link local concerns to global norms such as FPIC, to the emerging debates on national and transnational safeguards for REDD+, to donor safeguards and to human rights. The need for FPIC for all natural resource concessions is stipulated in Indonesia's REDD+ strategy. In addition, the recent constitutional court decision (*Keputusan* MK 35/PUU-X-2012) may further strengthen the rights of local and indigenous communities. DANIDA and the "Human Rights-Based Approach in German Development Policy" acknowledge the UN Declaration for the Rights of Indigenous Peoples and the right to FPIC (DANIDA, 2011: 2–3; Schielmann et al., 2013: 26). Peasants in Kunangan Jaya I explicitly stated that PT REKI failed to consult them adequately. This has been confirmed by other sources (Kreditanstalt für Wiederaufbau, 2012; Wirasapeotra and Octavian, 2012) and by activists of Yayasan CAPPA.[56]

PT REKI's failure to conduct FPIC in the settlements of Kunangan Jaya I provided an entry point for multi-scalar campaigns against the conservation company. With the support of CAPPA, peasants were able to lift the conflict to the provincial scale (e.g. the demonstration in front of the provincial parliament) and to the transnational scale (e.g. posting letters of complaint on REDD-Monitor, contacting PT REKI's donors). Multi-scalar resistance probably contributed to the at least partial acceptance of the land claims of the community of Kunangan Jaya I. In 2015, five years after the conservation company received its concession in Jambi, a new human–rights-based approach was announced by PT REKI. The new policy (Human Rights, Social and Community Engagement Commitment, HARSCEC) makes explicit reference to the right to FPIC for indigenous and local communities and stipulates social impact assessments (REKI, 2015). However, as mentioned above, announcing a commitment of five years after project implementation contradicts the very idea of FPIC, which is to provide sufficient information and allow communities to reject interventions. The extent to which these intended goals of the ILO 169 convention can be met in the context of a running project remains, at the least, questionable.

Social identity is an important factor in the conflict over Kunangan Jaya I. Global norms such as FPIC, the Cancun Safeguards and the decision of the Indonesian Constitutional Court have further increased the relevance of social identity and contributed to the construction of ethnicity as a powerful means to claim natural resources. The Provincial Forest Services, for example, have rejected the reclassification of Kunangan Jaya I to non-forest land, arguing that most of the settlers are not members of local ethnic groups and consequently do not have customary rights to the land. PT REKI's new HARSCEC policy, for instance, refers to FPIC for local and indigenous communities, not for peasant migrants. Ethnicity is a very dynamic category in the Harapan landscape. Interethnic marriages between Batin Sembilan and migrants are common and most of the migrants claim to have received land from local ethnic groups (Hein, 2013b; Hein et al., 2015; Hein and Faust, 2014). For the conflict between peasants and the oil palm company of PT Asiatic Persada, Steinebach (2013b: 73–74) has shown that becoming Batin Sembilan, e.g. through interethnic marriage but also by just claiming to be Batin Sembilan, is part of a set of strategies to reclaim land appropriated by the company (ibid.). Becoming Batin Sembilan seems to be a relatively inclusive and conflict-free process in the landscapes around the Harapan Rainforest. Migrants and Batin Sembilan actively cooperate, as the history of the formation of the settlements TSM and Tanjung Mandiri has shown (described in Chapter 3).

Finally, the conflict concerning Kunangan Jaya I confirms that local communities are heterogeneous. There is not a one and only "local community" that should be involved in conservation and development projects. This is not new information, especially not for human geographers and anthropologists, but it seems that many conservationists and development practitioners still neglect the complexity of social relations at the so-called "local or project level". Different actors e.g. peasants associated with Pak Kumis and the members of PERTAMA or the Batin Sembilan groups of Simpan Macan Dalam and Luar have different interests and are part of different networks. It is important to consider that not all community members were involved in conflict mediation. Ten community members were elected to represent the community in the initial negotiations. All community members were invited to regular meetings on the mediation process, stated a key informant.[57] However, while decisions were based on a majority vote of all community members not all villagers followed or were able to follow the process. The large distance between the dispersed settlements of Kunangan Jaya I and the lack of roads made it very complicated for community members to participate in community meetings.

We are here to stay: the conflicts in Camp Gunung and Tanjung Mandiri

The village-scale settlement initiatives Camp Gunung and Tanjung Mandiri were not spared from conflicts with the conservation company. In both settlements, indigeneity politics played a crucial role during settlement formation as well. As outlined in Chapter 3, both settlements were legitimized by the Batin Sembilan authorities and village governments (see also Mardiana, 2017: 73). A former member of the village government of Tanjung Lebar argued that the settlers in Tanjung Mandiri did not want to sign any agreements with PT REKI because the Batin Sembilan who owned the land had invited them to settle in the area, thus there was no need for a permit from PT REKI.[58] Additionally, he stated that the conservation and collaboration agreements with PT REKI would only provide access to land for a limited period. He further argued that this is not attractive for the settlers since they claim to have rights granted by the Batin Sembilan.

The conflict between the settlers of Tanjung Mandiri and PT REKI has not yet been resolved, negotiations on a potential collaboration agreement were unsuccessful. AMAN was initially involved in mapping indigenous land claims and facilitating negotiations between the community and PT REKI. According to a member of the village government in Bungku the agreement proposed by AMAN implied that settlers would have to stop oil palm cultivation, additional agricultural expansion, slash-and-burn agriculture and logging. This was rejected by the settlers. An AMAN activist in Jambi explained that some community members had doubts about the neutrality of AMAN. Consequently, AMAN stopped the mediation process. The activist also explained that the conditions for conflict mediation were very difficult because many different actors had competing interests. Some Batin Sembilan leaders were very heavily involved in land trade; some settlers and Batin Sembilan were rather poor peasants while other settlers resembled mafia-like speculators rather than peasants, thus "[. . .] it was difficult to find 'the right' people to negotiate with. Because we cannot talk with 200 people", the AMAN activist explained.[59]

According to a PT REKI project manager, a second mediation attempt facilitated by a number of state agencies failed after the Ramadan of 2013. The project manager argued that the negotiations failed because community representatives involved in the negotiations did not communicate intermediate agreements to all community members. The village government of Tanjung Lebar confirmed the view of AMAN that different community members had contrary opinions on how to deal with the conflict.

Settlers who I interviewed in 2013 described the situation as relatively calm: "We do not have an open or violent conflict with REKI". However, the situation changed in late summer 2013 when peasants destroyed cars belonging to the conservation company after rumors that the conservation company had destroyed oil palm plantings. The peasants in Tanjung Mandiri, in contrast to those in other settlements, were not interested in receiving support from NGOs. A settler explained, "We do not want NGOs involved, we do not want anarchy!"[60]

Especially in the Tanjung Mandiri area, rumors circulated that land conversion was not only driven by peasants but also by urban-based land speculators. PT REKI has argued that some of the ongoing land conversion was caused by urban residents from Bogor, Jakarta and Bandung. According to PT REKI, they are paying local peasants to convert land and to plant oil palms. These activities were difficult to verify and beyond the scope of this study. However, some Batin Sembilan complained that a number of actors have converted larger forest areas without requesting their permission. SPI members also complained about the presence of large land holdings within PT REKI's concession which have not been subject to prosecution under the forest policy. Land investments by urban-based citizens are relatively common in Jambi. Peasants and also urban residents met in Jambi city reported that higher-level civil servants frequently invest in rubber and oil plantations in the countryside, often in their regions of origin.

The conflict between the settlers of Camp Gunung and PT REKI intensified in 2010, after PT REKI and the armed BRIMOB police moved in and destroyed settlers' shacks. Settlers of Camp Gunung reported, as did their fellows in other parts of the conservation concession, that the conservation company did not conduct any appropriate form of community consultation before the destruction and eviction attempts took place. Some community members left the area during PT REKI's and BRIMOP operations and moved back later.

In 2011, after pre-negotiations, a participatory land tenure assessment was conducted and in 2012, as in the case of Kunangan Jaya I, the MoF offered small-scale forest concessions to the settlers. However, also as in Kunangan Jaya I, the MoF stepped back and called for a collaboration agreement between PT REKI and the settlers. In 2012, a few community members supported by the peasant organization STN organized a "long march to Jakarta" and protested in front of the MoF to attract additional attention to their situation. In 2013, during field research in the area, the settlers described the situation as rather calm and explained that PT REKI accepts their presence now. Finally, in 2017, representatives of Kunangan Jaya II and PT REKI signed an agreement permitting 30 households to use 135.31 ha of land (PT REKI, 2017).[61]

Peasants, migrants and the state: conflicts among state apparatuses concerning access to and control of the Berbak Carbon Initiative

The conflict about the transmigration site within the Berbak Carbon Initiative shows that different apparatuses of the state act in contradictive ways, reflecting different views and interests in society. The divergent interests of different actors within the state become visible in the forest reserve Sektitar Tanjung. Ongoing tensions between the state apparatuses aiming to facilitate rural development by promoting agricultural expansion and those aiming to foster conservation are inscribed in Berbak's forest frontier. The forest reserve Sektitar Tanjung is located approximately 4 km east of the main hamlet of the village of Kampung Laut (District of Muaro Jambi). In 2008, the Zoological Society London, the Jambi-based NGO Gita Buana, the Berbak National Park Agency and the Provincial Forest Agency agreed to use the political momentum after the Bali Climate Change Conference (COP 13) to expand protected areas in the surroundings of Berbak National Park. In the same year, the District Government of Muaro Jambi together with Javanese district governments started to establish a district-to-district transmigration program. One of the transmigration settlements was established east of Kampung Laut within the borders of the forest reserve Sekitar Tanjung, challenging the authority of the MoF and the integrity of the Berbak Carbon Initiative (Figure 5.4).

Conflict history and the formation of the transmigration settlement

In 2006, the district head of Muaro Jambi announced its interest in resettling Javanese farmers to Muaro Jambi to promote rural development. Two years later, the Transmigration Agency of Muaro Jambi (*Dinas Sosial, Tenaga Kerja dan Transmigrasi Kabupaten Muaro Jambi*) conducted an initial survey in Kampung

Figure 5.4 Picture of abandoned houses of the transmigration settlement of Kampung Laut
inside the Berbak Carbon Initiative

(Source: taken by the author, 2013)

Laut. By the end of 2008 the first houses of the settlements had been set up
(see Figure 3.2 for the location of the settlement). Only then did the district
head of Muaro Jambi (Bupati H. Burhanuddin) request a formal permit from
the Governor of Jambi for the conversion of state forest and for the establishment of a transmigration settlement (*Surat 522/776/KANHUT Tanggal 10
Desember 2008*) (WALHI Jambi, 2009). However, according to the existing legal
framework, neither the district government nor the provincial government has
the authority to permit the conversion of state forest; only the MoF has the
authority to do so. In 2009, without having received any de jure permit from
the forest authorities, the district head of Muaro Jambi announced that 150
transmigrants from Java would be resettled to Muaro Jambi in 2009 (Republika
Online, 2009). The agreement between the district of Muaro Jambi and three
East Javanese districts (Madiun, Lamongan and Blitar) permitted the relocation
of 225 households (ibid.). Of the first 150 transmigrant households, 75 would
be relocated to the village of Kampung Laut and 75 to the neighboring village
of Gedong Karya, as stated by the district head in a local newspaper (ibid.).
In Kampung Laut, peasants told me that one of the objectives of the settlement was to provide shelter for refugees displaced by the 2006 eruption of the

Sidoarjo mud volcano in East Java.[62] The eruption was probably caused by oil and gas drilling activities. The mudflows destroyed more than a dozen villages and agricultural land in the district of Sidoarjo (Mohsin, 2018). However, the three Javanese districts involved in the agreement were not affected by the mudflows and, when interviewed, staff of state agencies in Jambi were not aware of any link between the disaster and the district-to-district transmigration initiative. Nevertheless, this story was told among villagers. According to official district policies, the settlement in Kampung Laut was planned to provide land for settlers from East Java and for young, poor and landless Malay families from Kampung Laut. It was planned that each participant should receive 2 ha of agricultural land and 0.75 ha for a house and house garden. Furthermore, participants were told that they would receive further agricultural extension services.[63]

In 2009, the Provincial Forest Agency stopped the finalization of the settlement project, claiming that the project was being illegally conducted within the Forest Reserve Sekitar Tanjung, thus violating Forest Law 1999/41.[64] A few months earlier, the first local settlers from Kampung Laut had moved into their new houses in the transmigration settlement. After the intervention of the Provincial Forest Agency rumors emerged that the settlement had to be abolished. Consequently, most of the Malay households from Kampung Laut that had received land in the settlement sold their land to spontaneous migrants, mainly to second-generation transmigrants from the Batang Hari delta region. The Provincial Forest Agency only stopped the formal settlement formation process. The agency did not try to evict the remaining Malay settlers and did not try to stop the influx of spontaneous migrants. Staff of the Provincial Forest Agency argued that eviction by force would have exacerbated the situation and have led to chaos.[65]

Mediation between the different state actors involved in the process and community representatives from Kampung Laut has been long lasting. In 2013, after negotiations facilitated by the Governor of Jambi, it was announced that the conflict was solved. According to the community representatives, the MoF decided to release 150 ha of state forest land which had been mapped by representatives of the provincial parliament (*Dewan Perwakilan Rakyat Daerah*, DPRD).[66] This was confirmed by staff of the Provincial Forest Agency.[67] It was also announced by governmental officials (interviewed community members did not know which governmental agency made the announcement) that the first Javanese transmigrants would be relocated to Kampung Laut in 2014.[68] However, critical voices within the community argued that the Provincial Forest Service had not made a final decision on the settlement yet and that the negotiations with the central government in Jakarta continued.

In 2013, at least two contradictive conflict-solving processes were ongoing. The first foresaw the release of the settlement from forest land and the formation of a new and independent village. This approach was favored by the District Government of Muaro Jambi. At the same time, the Provincial

Forest Agency started to implement a conditional land tenure scheme, also involving the transmigrant community. The conditional land tenure scheme includes the allocation of forest land for the cultivation of jelutung and other forest species, aiming to rehabilitate the forest cover of the Forest Reserve (Figure 5.5).

In 2016, the development-oriented actors within the district and the village government seemed to have won the battle over the swamps of Kampung Laut. According to representatives of the village government, the transmigration settlement and 150 ha of agricultural land was officially released from the state forest territory.[69] A map of the National Land Agency (Figure 3.2D) from 2016 identifies the settlement as a "planned transmigration site" outside the state forest (Hein et al., 2018a). During a short visit I made in August 2016, the district agency of public works (Dinas Pekerjaan Umum) was constructing a new drainage channel and a dyke system to protect the settlement from annual flooding during the rainy season. In addition, the responsible neighborhood head told me that governmental agencies have promised that all houses and infrastructure of the settlement will be refitted, that all settlers will receive 1 ha of agricultural land, and that an oil palm company might invest in a new plantation estate involving the settlers.[70]

Figure 5.5 Degraded peat swamp inside the Berbak Carbon Initiative
(Source: taken by the author, 2013)

Development narratives, rent-seeking and new property

The establishment of the transmigration settlement of Kampung Laut not only challenged the MoF, but it also provided rent-seeking opportunities for village-scale and district-scale actors. The district head sought to legitimize and justify his initiative by referring to two well-known policy narratives of the new order era. He argued that Javanese transmigrants would act as model farmers, augmenting the agrarian potential of the region. Secondly, he referred to rural development and economic growth to justify forest conversion. Muaro Jambi's long-term development plan (*RPJPD, Rencana Pembangunan Jangka Panjang Daerah 2006–2025*) highlights the relevance of the transmigration program for developing intensive agricultural systems (Pemerintah Kabupaten Muaro Jambi, 2012: (2) 157).

Detailed information on the intentions of the district government in establishing the settlement beyond the references to rural development and economic growth was not available. However, it is worthwhile mentioning that the circumstances of the establishment of the transmigration settlement were investigated by the Jambi police following accusations of corruption against staff members of the district's transmigration agency and the construction company involved (JambiekspressNews, 2014).[71] Furthermore, the formation of the settlement and the allocation of land to local famers probably provided political rents (e.g. popularity and votes in district elections for the district head and his allies). Interviewed community members in Kampung Laut described the district head, H. Burhanuddin, as very supportive of the village community. He visited the village three times and talked directly to community members. Moreover, during one of his visits he permitted the first local transmigrants to move into the transmigration settlement prior to the official finalization of the settlement. In addition, it is important to consider that H. Burhaniddin seems to be an active advocate for the land rights of peasants within the state forest. As described in Chapter 3, he also supported the community of the settlement of Tanjung Mandiri in the Harapan Rainforest. He legitimized their presence within the PT REKI's conservation concession by celebrating the annual rice harvest ceremony (*Panen Raya*) in the settlement and promised that the Tanjung Mandiri would be released out of the Harapan Rainforest Project (without having the de jure competency to do this).

The transmigration settlement provided significant benefits for the village government of Kampung Laut as well, in particular, because the village government, especially the neighborhood heads (*Ketua Rukun Tetangga (Ketua RT)*), were in charge of selecting the local participants. In interviews project participants mentioned that only poor, landless and married couples were able to participate. However, in informal conversations key informants stated that villagers who did not fit into these categories also received land within the settlement. Consequently, the formation of the settlement provided room for rent-seeking at the village scale. The authority of the village government to select local participants to allocate land within the state forest thus provided significant benefits for the village elite. Moreover, the conflict transformed the transmigration settlement in two different ways: first, from a district-backed initiative intended to provide land for Javanese migrants and local households to a destination for spontaneous migrants; second, local project participants transformed themselves into land traders in order to cope with the conflictive situation. In 2013, 38 households lived in the settlement, only five of them were initial local project participants from Kampung Laut. All other local participants have sold their land. Official transmigrants from East Java never arrived in the settlement according to key informants in Kampung Laut.[72]

The conflict over access to and control of the forest reserve altered access and property relations. The conflict provided local Malay elites with the ability to expand their village territory and thus to create additional income through selling land to spontaneous migrants. The spontaneous migrants became new land-owners holding property rights legitimized by the local Malay and indirectly by the district head of Muaro Jambi. At the same time new types of properties have emerged. The conditional land-tenure schemes introduced as a conflict-solving mechanism by the Provincial Forest Agency, for instance, provide land-use rights bound to conservation regulations.

Competing state apparatuses and implications for REDD+

Settlers and the village community of Kampung Laut experienced the conflict mostly as a conflict between different state apparatuses. Key informants stated, for instance, that the Transmigration Agency was not aware of the boundaries of the Forest Reserve and complained that the Provincial Forest Agency inter-vened only after construction of the settlement started.[73] Staff members of the Provincial Forest Agency challenge the legitimacy of the scheme and argue that the transmigration settlement as such is not an "official scheme".[74] Gita Buana, the local NGO involved in the Berbak Carbon Initiative highlighted that the settlement further increases the risk of disturbance in the National Park. Staff of Gita Buana wondered about the contradictive policies of different state appa-ratuses.[75] Here, the implementation of the Berbak Carbon Initiative has added an additional conflict layer; this has also been recognized by some community members, as indicated by statements such as, "[. . .] the forest reserve is owned by the world".[76]

The agreement between the Zoological Society of London (ZSL) and the Provincial Forest Agency states that the forest reserve could supply carbon credits with a value of more than 200 million USD within 30 years (Dinas Kehutanan Provinsi Jambi, 2013). The exact value would, of course, depend on carbon prices on the voluntary market or on future binding carbon markets. Even more importantly, the production of carbon credits requires the integrity of the project boundaries and especially the effective avoidance of deforestation, forest fires and further peat conversion. However, precisely these objectives are those challenged by the district government through the implementation of the transmigration settlement (Hein et al., 2018a). The case shows that apparently local conflicts between different state actors at Berbak's forest frontier might challenge the production of forest carbon credits and the permanence of avoided greenhouse gas emissions. Consequently, they become globally relevant. To produce carbon credits the seller (e.g. the project developer, in this case ZSL and partners) has to guarantee that the forest remains intact – at least for a certain period. The buyer of the carbon credits uses the credits to compensate for greenhouse gases that have already been emitted (in other words, to offset). If conflict between different state apparatuses challenges the permanence of the avoided greenhouse gas emissions (e.g. because of forest conversion), the offsetting would be undermined and this could create conflicts between the local sellers and the transnational buyer of carbon credits.

Summary and preliminary conclusion

REDD+ and the idea of forest carbon offsetting have transnationalized apparently local agrarian conflicts in at least three different ways. First, the UNFCCC REDD+ governance framework, international conventions and declarations on the rights of indigenous people, and transnational REDD+ safeguards provide transnational norms and soft laws that peasants and indigenous groups increasingly use to campaign for and to defend access to natural resources and land. Second, the contestations involve transnational actor coalitions and conflicts occur on multiple scales. Conservation and REDD+ projects are often implemented by a wide network of transnational actors including state agencies, donors, private companies and environmental organizations. Protest and resistance against green enclosures, as the conflicts described in this chapter show, have an increasingly transnational character, linking peasants with global peasant organizations, advocacy NGOs and the private donors, such as Misereor and the Ford Foundation, who support these groups financially. Third, the very idea of offsetting changes the meanings of agrarian conflicts. Forest carbon offsetting transfers responsibility for reducing greenhouse gas emissions from global centers to the periphery and in most cases to regions that have contributed much less to global warming than the urban centers in the Global North. Peasants raised concerns that green enclosures challenge their livelihood strategies, asking whether their lives are not more important than reducing greenhouse gas emissions. In this regard, the struggles in Jambi are also struggles for climate justice. The idea that REDD+ could support a sustainable transition in rural areas (e.g. Murdiyarso et al., 2012: 680) did not gain much traction in the landscapes around the Berbak Carbon Initiative and the Harapan Rainforest. For many peasants, as well as for conservationists, trade-off thinking dominated: conservation versus rural development. Only recently, in the context of conflict mediation, has it seemed possible that conditional land tenure for rubber agroforestry might gain traction as a policy to align mitigation objectives with local development.

The conflicts about access to and control of the Harapan Rainforest cases show that peasants supported by NGOs were able to defend their land claims and have challenged the commodification of forest carbon. Successful peasant resistance relied on scale jumping and on the construction of scales of meaning and regulation. Scales shape resistance opportunities, permitting, for instance, scale jumping to pre-existing higher scales and alliances with actors that are able to raise protests at higher scales. At the same time resistance activities can also change existing scales (e.g. expansion of village scale of regulation) or produce new scales (e.g. SPI territory). The conflicts about the Berbak Carbon Initiative case (and, to a lesser extent, the Harapan Rainforest case, as well) show that different apparatuses of the state follow different interests reflecting different interest groups within society. The conflicts in both projects indicate that the rescaling of access and property relations, in other words the spatial expansion or reduction of regulation and authority, is not a conflict-free process. Scales of regulation reflect power relations between different authorities, they are maintained and reproduced by actors requesting land titles and they are challenged by other actors requesting land titles from a competing authority. The scales of tenure regulations in place represent a fragile stand-off or compromise between different public authorities, including customary leaders and various state apparatuses.

Furthermore, the illustrated cases show that rescaling has provided extra leeway for successful peasant resistance. Especially REDD+ attracted attention to structural inequality at the forest margins. International norms such as FPIC and the Cancun Safeguards strengthened the position of peasants and especially of indigenous groups. As the Harapan Rainforest case illustrates, the failure to conduct FPIC provided entry points for transnational resistance campaigns. Peasants used different multi-scalar resistance strategies. They built alliances with customary leaders, village governments and with NGOs and transnational peasant organizations, such as La Via Campesina. Especially those peasants affiliated with SPI were able to raise concerns about the Harapan Rainforest project at the global scale. Peasants supported by CAPPA were able to send protest notes to donors and to organize protest at the provincial scale. Nevertheless, it was not only scale relations that proved relevant for successful resistance (Hein et al., 2016: 387); network relations, e.g. linking up with indigenous groups or cooperation among local NGOs, were shown also to be of particular importance for successful peasant resistance (ibid.).

The ability to resist against PT REKI's land claims, or – from PT REKI's perspective – their defense of the integrity of the Harapan Rainforest, is also a matter of power. According to Gaventa (1982: 23), rebellion or the challenge of power "[. . .] may develop if there is a shift in the power relationships – either owing to loss in power of A or gain in power of B". In the first place, the changed power relations after the fall of Suharto provided the opportunity to organize and transform hidden resistance into larger, open resistance that actively challenged hegemony. Following Gaventa (2006: 31), I have argued that strategic alliances and multi-scalar strategies paved the way for successful land occupations and later on for the successful resistance of the peasants of Kunangan Jaya I and of SPI. The ability to form alliances with actors on different scales and to construct political campaigns tailored for different scales (e.g. land distribution at the local scale, agrarian reform at the national scale, and anti-REDD+ at the global scale) contributed to their success. Peasant organizations became a relevant force in Jambi. SPI, for instance, has developed visible power. The organization has 15,000–20,000 members in Jambi and financial resources from membership fees.[77] Yayasan CAPPA, as the main NGO supporting peasants in Kunangan Jaya I, relies on a transnational network and has received significant support from donors (e.g. Misereor and the Ford Foundation). Yayasan CAPPA has the capacity to organize protest and to act as an up- and downward translator. The organizational strength of SPI and the capacities of Yayasan CAPPA have contributed to the ability of the different peasant groups to resist PT REKI, the MoF and the mobile police brigades.

Notes

1 Interview with an AMAN activist, Jakarta, 27.07.2012.
2 Interview with an AMAN activist, Jakarta, 27.07.2012.
3 Interview with an AMAN activist, Jambi, 02.08.2013.
4 The internal administration of SPI corresponds to the hierarchical Indonesian administrative system. In villages with more than 25 SPI members the organization establishes a basis. A basis is the smallest organizational unit led by an elected head. The elected head (*ketua basis*) represents the basis at the *ranting*, which is a roundtable where all *ketua basis* of the sub-district meet. The elected head of the *ranting* represents the organization at the district level. Central decision-making bodies are the regional and national conferences. Decisions made at the national level are communicated as recommendations to the lower levels.
5 Interview with an SPI activist in Jakarta, 20.06.2013 and with SPI activist in Jambi, 12.07.2013.
6 Interview with a key informant in Bungku, 13.09.2012.
7 Dragon blood, *Daemonorops draco,* is a specific rattan species that produces a red resin.
8 Interview with a key informant in Bungku, 12.09.2012.
9 RSPO is a transnational certification body that promotes sustainable oil palm plantations.
10 Interview with a key informant in Tanjung Lebar, 27.07.2013.
11 Interview with key informants in Tanjung Lebar, 25.07.2013.
12 Interviews with key informants in Tanjung Lebar, 21.07.2013.
13 Interview with an SPI activist in Jambi, 12.07.2013.
14 Interview with a staff member of Burung Indonesia, 11.10.2013.
15 Interviews with key informants in Tanjung Lebar, 21.07.2013, 22.07. 2013, with an SPI activist in Jambi, 12.07.2013, and with KfW staff members in Frankfurt am Main, 04.02.2014.
16 Interview with an SPI activist in Jakarta, 20.06.2013.
17 Interview with a staff member of PT REKI in Bungku, 30.07.2013 and with staff members of KfW, Frankfurt am Main, 04.02.2014.
18 Interview with an SPI activist in Jambi, 12.07.2013 and with KfW staff in Frankfurt am Main, 04.02.2014.
19 Interview with KfW staff members in Frankfurt am Main, 04.02.2014.
20 Interview with a KfW staff member in Frankfurt am Main, 19.04.2012.
21 Interview with an activist of AGRA, Jambi, 18.09.2013.
22 Interview with staff members of KfW, Frankfurt am Main, 04.02.2014.
23 Interview with staff members of KfW, Frankfurt am Main, 04.02.2014.
24 Interview with key informants in Tanjung Lebar, 22.07.2013.

25 Interview with an SPI activist in Jakarta, 20.06.2013.
26 Interview with an SPI activist in Jambi, 12.07.2013.
27 Interview with a key informant in Tanjung Lebar, 21.07.2013.
28 Interview with a key informant in Tanjung Lebar, 22.07.2013.
29 Interview with a staff member of Carbon Synthesis, 11.10.2012.
30 Interview with a staff member of PT REKI in Bungku, 30.07.2013.
31 Interview with a staff member of PT REKI in Bungku, 30.07.2013.
32 Interview with a key informant in Tanjung Lebar, 21.07.2013.
33 Interview with a staff member of PT REKI, Bogor, 11.10.2013.
34 Interview with a key informant in Bungku, 23.08.2013.
35 Interview with key informants in Bungku, 10.07.2013 and 25.08.2013.
36 Interview with key informants in Bungku, 12.09.2012.
37 Interview with key informants in Bungku, 12.09. 2012 and 10.07.2012.
38 Interview with a key informant in Bungku, 23.08.2013.
39 Interviews with key informants in Bungku, 10.07.2013, 23.08.2013 and 24.08.2013
40 Interview with a key informant in Bungku, 24.08.2013.
41 Interview with a key informant in Bungku, 10.07.2013.
42 Interview with a key informant in Bungku, 25.08.2013.
43 Interview with a key informant in Bungku, 23.08.2013.
44 Interview with a staff member of PT REKI in Bungku, 30.07.2013.
45 Interview with an SETARA activist in Jambi, 18.07.2013.
46 Interview with staff of KfW in Frankfurt am Main, 04.02.2014.
47 For the reclassification of forest land to non-forest land, specific criteria have to be met. The criteria formulated by the General Director of Forest Planning in 1994 consist of biophysical criteria (e.g. soil type and altitude) and socio-economic criteria (e.g. land use, ethnicity of settlers, proof of land ownership).
48 The term refers to Suku Anak Dalam, a post-colonial and deprecatory term for indigenous communities, in this case the term refers to the Batin Sembilan.
49 Batak is an ethnic group from the Province of North Sumatra; Medan is the capital of North Sumatra.
50 Interview with a staff member of Dinas Kehutanan Provinsi Jambi, Jambi, 19.09.2012.
51 Interview with a staff member of PT REKI in Bungku, 30.07.2013.
52 Interview with a staff member of PT REKI in Bungku, 30.07.2013.
53 KM 35 is a small settlement of Kunangan Jaya I located on a former logging road between Simpang Macan Luar and the airfield of PT REKI.
54 Interview with a key informant in Bungku, 12.09.2012.
55 Interview with a key informant in Bungku, 12.09.2012.
56 Interviews with a CAPPA activist in Jambi, 18.07.2013.
57 Interview with a key informant in Bungku, 25.08.2013.
58 Interview with a key informant in Tanjung Lebar, 26.07.2013.
59 Interview with an AMAN activist in Jambi, 02.08.2013.
60 Interview with a key informant in Tanjung Lebar, 27.07.2013.
61 Interview with key informants in Bungku, 09.07.2013.
62 Interview with key informants in Kampung Laut, 30.08 2013 and 09.08.2016.
63 Interviews with key informants in Kampung Laut, 29.08.2013 and 30.08.2013.
64 Interview with a staff member of the sub-district administration (Kecamatan) of Kumpeh in SuaKandis, 02.09.2013. Interviews with key informants in Kampung Laut, 29.08.2013 and 30.08.2013 and with a staff member of the Dinas Kehutanan Provinsi of Jambi in Jambi, 27.08.2013.
65 Interview with a staff member of Dinas Kehutanan Provinsi of Jambi in Jambi, 27.08.2013.
66 Interviews with key informants in Kampung Laut, 29.08.2013 and 30.08.2013.
67 Interview with a staff member of the Dinas Kehutanan Provinsi of Jambi in Jambi, 27.08.2013.

68 Interviews with key informants in Kampung Laut, 29.08.2013 and 30.08.2013.
69 Interview with key informants in Kampung Laut, 09.08.2016.
70 Interview with a key informant in Kampung Laut, 09.08.2016.
71 Interview with a staff member of Dinas Kehutanan Provinsi of Jambi in Jambi, 27.08.2013.
72 Interview with a key informant in Kampung Laut, 29.08.2013.
73 Interview with key informants in Kampung Laut, 29.08.2013 and 30.08.2013.
74 Interview with a staff member of the Dinas Kehutanan Provinsi, 27.08.2013.
75 Interview with a staff member of Gita Buana, 22.08.2013.
76 Interview with a key informant in Kampung Laut, 29.08.2013.
77 Interviews with SPI activists in Jakarta, 20.06.2013 and Jambi, 12.07.2013 and with key informants in Tanjung Lebar, 22.07.2013.

6 Conclusion

Towards a political ecology of transnational agrarian conflicts

This book has provided a comprehensive analysis of agrarian conflicts in the context of REDD+ implementation, forest carbon offsetting and privatized conservation. The incidents in Jambi illustrate that transnational conservation initiatives do not act in a social and political vacuum. They show that the ability to define "nature" as a space for shifting cultivators, as oil palm plantation estates or as forest carbon offset is linked to power relations. My first step towards drawing conclusions is to sum up a number of key arguments and key results. Using the agrarian conflicts in Jambi as a point of departure, I then propose elements for a political ecology of transnational agrarian conflicts that explicitly considers the role of socio-spatial relations and of power and property for future research in Indonesia and beyond.

At the beginning of the book, I raised two arguments which help to explain the complexity of the socially differentiated impacts of REDD+. First, I argued that REDD+ can be many different things at different political scales and for different actors and their respective storylines and discourses. Conservationists and economists have conceptualized REDD+ as a win–win mechanism. However, conflicts in Jambi and in other parts of the world (e.g. for Africa: Leach and Scoones, 2015) show that instead of providing win–win outcomes, the mechanism has rather maintained and reproduced existing inequalities and power asymmetries. REDD+, as most other conservation instruments, restricts access to land and land use. In Sumatra, as in other world regions, state territorialization projects have allocated large amounts of land to corporate actors for the sake of development and neglected indigenous and peasant rights to land. REDD+ has thus created additional threats for local livelihoods instead of facilitating a just transition to sustainability as intended by some (Grieg-Gran et al., 2015; Holmgren, 2013).

REDD+ in Jambi, as outlined in Chapter 4, is neither a single policy nor a coherent set of conservation interventions. REDD+ in Jambi takes many different forms. The whole province is a REDD+ pilot with a detailed province-wide action plan for reducing deforestation, including policies to solve existing land tenure conflicts. However, most of the proposed policies have remained on paper and exceed the formal competencies of the provincial government. In parallel, a number of independent, often non-state actor-driven and donor-financed forest conservation initiatives have been implemented. Some of them aim to sell carbon credits on voluntary carbon markets, others form part of voluntary commitments to reduce greenhouse gas emissions and support biodiversity conservation by bilateral and private donors such as Germany, Denmark, the UK and Singapore Airlines. The population affected by these conservation initiatives calls them "lungs of the earth" or "carbon toilet" and asks rather critically, "Why did the rich countries buy the oxygen in our forest?" Last but not least, REDD+ has facilitated the diffusion of international norms such as free, prior and informed consent (FPIC). The explicit acknowledgement of indigenous and local communities in UNFCCC decisions on REDD+ has provided additional legitimacy for ethnicity-based land claims. Yet, the new norms have not been translated into new de-jure rights and have not been acknowledged by all conservationists, as the Harapan Rainforest case shows. However, they provide entry points for organizing transnational resistance and they attract additional attention towards structural inequality at the forest margins.

Second, and probably more fundamentally, I drew on work by Fairhead and colleagues (2012) to argue that, in the context of forest carbon offsetting and market-based conservation, "[. . .] the basic questions of the agrarian political economy are as relevant as ever [. . .]". Land is at the heart of the agrarian question. The agrarian conflict and peasant resistance described in Chapters 3 and 5 of this book are first and foremost conflicts about access to and control of land, which is the most important means of production in landscapes dominated by agriculture. Most of the conflicts about access to and control of land in Jambi are older than the recent conservation interventions. They are rather caused by historically contingent structural inequality and by the non-recognition of customary rights than by conservation interventions as such (c.f. Hein and Faust, 2014: 25). However, the historical conditions that facilitate the implementation of new conservation areas and REDD+ are neither primordial nor unchangeable. They were established by the colonial and post-colonial state to strengthen territorial control and to facilitate the extraction of nature resources and the expansion of large-scale agricultural estates. Today, exactly these conditions are used and reproduced by conservation actors for the production of a new commodity, namely forest carbon, but also for rather fictitious commodities referred to as the "currencies of conservation spectacle" by Igoe and colleagues (2010: 494), namely new "successful" protected areas and spectacular images of flagship species such as the Sumatran Tiger and Sumatran Elephant.

The access and property relations in landscapes discussed in this book are complex, conflictive and highly dynamic, challenging the expansion of protected areas. They are the outcome of competing property rights that have evolved across time and that are regulated on different scales and issued by competing authorities. I argue that property rights, the notion of a state and especially scales of regulation are inseparably linked. Colonial scales of meaning and regulation facilitated access to land and natural resources to extract benefits for Dutch colonial authorities and corporate actors, challenging the pre-existing scalar arrangements of customary communities. The autocratic Suharto regime established a national scale of meaning based on moderniza-tion and development narratives; and implemented a complementary national scale of regulation facilitating access to land for timber and oil palm companies. In the late 1990s, political struggle induced far-reaching scalar restructuring and changing dialectical relationships between structure and agency, further supported by the objective of reallocating land and natural resources. Decen-tralization and changed power constellations have significantly widened the opportunities for local actors to access land, as the informal settlements within the Harapan Rainforest indicate. The diffusion of international norms and transnational carbon standards in the context of REDD+ can be considered as the most recent, still ongoing rescaling process. Peasant organizations such as La Via Campesina and SPI organized protests at UN conferences. Consequently, successful peasant resistance further reproduced the emerging global scale of REDD+ governance.

Elements for a political ecology of transnational agrarian conflict

In the following section different explanations and interpretations of the main results of this study will be outlined as a starting point for a political ecology of transnational agrarian conflict, thus for future empirical research. Explanations and interpretations will be mainly based on the core analytical categories out-lined in Chapter 2, namely scale, power, territory and property.

Socio-spatial relations and conflictive modes of production

The different ways land is used influences scalar arrangements and property rela-tions (Zulu, 2009: 690). I consider conservation including REDD+, corporate oil palm cultivation, smallholder oil palm and rubber cultivation, and shifting cultivation as mutually exclusive modes of production which create different socio-spatial arrangements and competing access and property relations.

Shifting cultivation, hunting and gathering practices were important modes of production in Jambi until the 1980s. Today, they have become less relevant in Jambi as well as in other frontier landscapes in the Global South (Hall et al., 2011: 106; Li, 2002: 421; Mertz, 2009). Shifting cultivation within and around the Harapan Rainforest is deeply entangled with a pre-existing watershed- and lineage-based socio-spatial organization. Shifting cultivation and also hunting and gathering practices require vast areas of land controlled by lineage and sub-lineage leaders. The corporate and smallholder expansion of oil palm and rubber cultivation and the expansion of conservation areas have significantly limited the land available for shifting cultivators. At the same time and conse-quently, modes of production access, property relations, and entangled scalar arrangements have changed. Watershed and lineage scales have been replaced by village, district, national and transnational scales that particularly facilitate oil palm cultivation and conservation as new modes of production. However, these changes – as outlined in the previous chapters – were highly conflictive.

The introduction of rubber by Dutch colonial authorities, the promotion of oil palm cultivation by the Indonesian government, and the reputation of oil palm as a modern crop (Hein et al., 2016; Locher-Scholten, 2004; Schwarze et al., 2015) incentivized peasant farmers to convert former swiddens to permanent rubber and oil palm plantations[1] (Schwarze et al., 2015: 1). Local communities cultivate oil palm even within the Harapan Rainforest and, to a lesser extent, within the Berbak Carbon Initiative. In many cases, the expansion of smallholder oil palm and rubber plantations has been promoted by customary, village and district authorities. As in other parts of Indonesia, they consider the conversion of state forest as a legitimate response to colonial and post-colonial dispossessions (Lukas, 2014; Peluso et al., 2008; Tuong, 2009). Land allocation and forest conversion has been facilitated by village governments and customary leaders. To provide a minimum of tenure security for peasants cultivating oil palm and to permit land trade, village governments started to issue different types of land titles, thus establishing a village scale of land tenure regulation. By requesting village-scale land titles, peasants reproduced and maintained the new scale of regulation. Moreover, I argue that the new village scale of land tenure regulation reflects and supports the requirements of a specific mode of production: smallholder oil palm cultivation. Village-scale land titles have facilitated the expansion of smallholder oil palm plantings, the commodification of former family or lineage-based property and the formation of a land market within the conservation areas studied.

Corporate oil palm cultivation has transformed rural Indonesia significantly in recent years. More than 10 million hectares of land are used for oil palm cultivation in Indonesia, approximately two-thirds by large-scale corporate oil palm plantation estates. Since decentralization, access to plantation concessions (HGU) has been mainly regulated at the district scale. Commercial oil palm plantations require large land holdings, full land control and labor (Hall et al., 2011: 92). Oil palm companies are usually well-positioned in regard to the three power dimensions developed by Lukes (2005) and Gaventa (1982). In consequence, they have been quite successful in accessing land rights for vast areas. Especially around the Harapan Rainforest, oil palm concessions overlap with land claimed by local and indigenous communities. Previous modes of production, such as logging, still permitted the co-existence of hunting and gathering and, in some cases, even of shifting cultivation. In contrast, commercial large-scale oil palm cultivation requires full land control and does not permit any co-existing land-use practices. In consequence, the introduction of commercial large-scale oil palm cultivation led to displacements and to conflicts with other right holders involved in other modes of production.

Another mode of production that requires large land resources is conservation and REDD+. The extent to which conservation and REDD+ can be fully considered as modes of production is the subject of heated debates. Political ecologists in particular consider them as being intrinsically linked to specific modes and relations of production related to the commodification of ecosystem services (Brockington and Scholfield, 2011; Castree, 2003; Corbera and Brown, 2010; Escobar, 1996; Fairhead et al., 2012; Garland, 2008; Kelly, 2011; McAfee, 1999; Zimmerer, 2000). In contrast, some conservationists consider their interventions rather as a counter movement aiming to protect the non-commodity status of wildness and to hinder the commodification of forest products such as timber (Li, 2008: 125). More market-oriented conservationists believe that only the full commodification of nature would help to protect increasingly shrinking habitats for non-human species (Igoe et al., 2010).

PT REKI and actors involved in the Berbak Carbon Initiative both support specific so-called biodiversity-friendly and low-carbon land use practices. But, like many other conservation initiatives, they also produce (or plan to produce) specific "fictitious commodities" (Nevins and Peluso, 2008: 19) as well as specific tradable commodities, such as sustainable products, ecotourism and forest carbon credits. Fictitious commodities, in other words symbolic commodities, produced by the conservation company PT REKI in the Harapan Rainforest include biodiversity hotspots, the last remaining patches of lowland rainforests of global importance, and a refuge for the indigenous Batin Sembilan. Tradable commodities produced by PT REKI include ecotourism and sustainably produced non-timber forest products (REKI 2014). As other economic actors, conservation NGOs and companies have to show success and highlight the economic advantages of their activities (Igoe et al., 2010: 496) vis-à-vis potential competitors. The RSPB and the ZSL advertise their projects with pictures of the endangered Sumatran tigers to illustrate that donations would help to save these species. Both projects, despite all the conflicts and controversies, merchandise their projects as success cases. ZSL (2015), for instance, has just changed the time line of the project in its most recent brochure on the Berbak Carbon Initiative, concealing that the project has made only limited progress since 2013. The NABU (2014), one of PT REKI's German partners, presents the Harapan Rainforest project as a success case and failed to even mention the ongoing conflicts in an article on their webpage.

Processes of the (fictitious) commodification of nature and new "green" modes of production, such as offsetting, are elements of the ecological phase of capitalism (Escobar, 1996: 326), shaping nature and society (Brockington et al., 2008: 5). In the landscapes around the Harapan Rainforest and Berbak Carbon Initiative, they encounter pre-existing and competing modes of production (especially smallholder and corporate oil palm cultivation) associated with different scalar arrangements or, in the terms used by Karl S. Zimmerer and Thomas J. Bassett (2003b: 288), with different geographies of resource access. The establishment of conservation projects involves the containing of space, thus the construction of conservation territories that fix social relations of conflict and cooperation (Zimmerer, 2000: 360). The formation of the Harapan Rainforest and Berbak Carbon conservation scales has changed the meaning of the contained space and has transformed them to objects of environmental governance (Cohen and McCarthy, 2014: 2). They have been produced by rescaling to a physical space for which no direct state authority or jurisdiction existed a priori (ibid.).

It is important to highlight that the state and its different apparatuses have a key role in the aforementioned socio-spatial processes. A number of subsequent state interventions have facilitated the expansion of capitalist modes of production which have changed access and property relations (Beckert et al., 2014; Brad et al., 2015; Fold and Hirsch, 2009; Faust, 2007; Hein et al., 2016; Nevins and Peluso, 2008). The formation of the specific scalar arrangements facilitating privatized conservation, corporate oil palm and smallholder oil palm cultivation have been actively promoted by state actors responding to discourses of market environmentalism and development. Conflictive access and property relations and conflictive scalar arrangements, as discussed above, can also be considered as conflicts between different apparatuses of the state over societal relationships with nature. The promotion of settlement and agricultural production within the Harapan Rainforest by village and district governments is one example. The transmigration settlement as a means to promote rural development within the Berbak Carbon Initiative is an even more astonishing case. Both cases indicate that conflicts between different state apparatuses, especially between conservation and development agencies (Brand et al., 2011: 150; Hein et al., 2018a), are additional explanations for conflictive access and property relations.

Rescaling

As described throughout the book, rescaling caused by state transformation is another relevant explanation for changing and conflictive access and property relations. Following Reed and Bruyneel (2010), I have identified three relevant scalar processes caused by state transformation processes, namely up-scaling towards transnational or international state apparatuses, down-scaling towards sub-national state apparatuses, and scaling-out towards non-state actors (e.g. conservation companies). State transformation from the colonial state, via the interventionist development state to the decentralized national competition state is a main cause of scalar restructuring and rescaling in Indonesia. In other words, up-, down- and out-scaling have been caused by regime change and policy shifts.

Up-scaling is taking place in the context of REDD+. The trading of REDD+ credits and multilateral or bilateral result-based payments for emission reductions require homogenous rules governing carbon accounting but also governing the acknowledgement of community rights and the design of participation processes. REDD+ has not yet led to a coherent scale of global forest governance. The so-called Cancun Safeguards of the UNFCCC and transnational carbon standards such as the Climate, Change, Community and Biodiversity Standard (CCBS) indicate that specific rights for local and indigenous communities have been up-scaled to the emerging global scale of forest governance. Yet the application of global norms governing community involvement (such as free, prior and informed consent) is still contested in many countries implementing REDD+ policies (Delgado-Pugley, 2013; Díaz, 2014; McCarthy, 2012). Many local and indigenous communities living within or adjacent to REDD+ pilot projects have not been involved in project planning, have not been informed about REDD+ and have not been asked to give their consent (Hein and Garrelts, 2014: 326; Mcculloch, 2010; Zelli et al., 2014: 104–106).

The down-scaling of land and natural resource governance in Indonesia has mainly taken place as a result of decentralization after the fall of Suharto. Indonesia's big bang decentralization can be considered a response to changing power relations and as a process that then further changed power relations. The decision to decentralize Indonesia was taken in the context of a relatively weak central state and of reappearing separatist tendencies (Hofman and Kaiser, 2002: 2). As explained in Chapter 3, the decentralization policies have only led to limited permanent de jure change of Indonesia's forest governance; however de facto changes have been significant as the informal settlements within the Harapan Rainforest, village-scale land titles and numerous studies on land tenure and conflict across the Archipelago indicate (Adiwibowo, 2005; Galudra et al., 2011; Lukas, 2014; Peluso et al., 2008). Down-scaling has increased the power of local state apparatuses and local and customary elites at the expense of national apparatuses. Within and around the Harapan Rainforest and the Berbak Carbon Initiative, power gains (e.g. mainly visible power) have been mainly absorbed by local elites, indicating that more decentralized or customary forms of environmental governance arrangements are not necessarily more favorable for social equality than more centralized ones. Studies focusing on the impacts of decentralization on access to land and forest resources on Indonesia and beyond indicate similar patterns of elite capture (McCarthy, 2004; Resosudarmo, 2004; Larson and Soto, 2008).

Scaling-out refers to the delegation of former state functions to non-state actors (Cohen and McCarthy, 2014: 13–14; Reed and Bruyneel, 2010: 648). Especially in the Americas, non-state actors have played a greater role in conservation since the 1990s through their involvement in payment for ecosystem service schemes (PES) and conservation concessions (Ellison, 2003; Langholz et al., 2000; Wolman, 2004). Indonesia started relatively late to privatize conservation and implement PES (Heyde et al., 2012: 1). It was the 2000s before the Indonesian government introduced privately managed ecosystem restoration concessions after successful lobbying by PT REKI's shareholder Burung Indonesia. For the very first time the concession delegated protected-area management within the state forest to non-state actors. Holding an ecosystem restoration concession provided PT REKI with the de facto authority to take over other state functions such as law enforcement and the legitimation of property rights. By scaling-out the management of the Meranti-River-Kapuas-River forest block to PT REKI, the Indonesian MoF delegated a whole bundle of rights to the conservation company, including the right to exclude other land users, in particular peasant migrants.

Agency, power and the production of scale

Neil Smith and other scholars noted that the production of scale is not only the outcome of modes of production and state interventions but also of human agency and social and cultural practices (Marston, 2000; Smith, 1992). The conflicts in Jambi confirm especially the relevance of agency for scale production. Peasant and indigenous rights movements used the extended room for maneuver of the *Reformasi* era to organize multi-scalar resistance campaigns against the land and forest allocation policies of various state apparatuses. In addition, they constructed alternative scales of meaning and regulations to legitimize forest conversion and land allocation, as shown by the different settlement and forest conversion projects organized by specific individuals supported by village governments and peasant movements. SPI and AMAN have challenged existing scalar arrangements through up-scaling resistance to the national and transnational scales.

Political scales are spatial delimitations of political power, argues James Meadowcroft (2002: 170). In consequence, changing power relations are an important explanation for rescaling. The above-mentioned resistance activities challenged power relations and scalar structures. Initially regime change, democratic freedom and subsequent decentralization decreased the power of the central state, providing the opportunity to transform hidden resistance into open resistance and actively challenging hegemonic actors such as the MoF. Gaventa (1982: 24) argues that in order to change power constellations in a specific arena, "the powerless" have to challenge all three dimensions of power. Taking again the example of the Harapan Rainforest, actors such as SPI were able to challenge *invisible* power by formulating political aims (e.g. implementation of an agrarian reform based on the Basic Agrarian Law, environmental justice), *hidden* power by mobilizing peasants and political allies (e.g. migrants in search for land and customary leaders), and *visible* power by developing material resources and organizational strength (e.g. land resources and members). Visible power permitted SPI to engage in open conflict with PT REKI and the MoF. SPI challenged the MoF and PT REKI on different scales and constructed and successfully defended their settlements within the Harapan Rainforest as a new scale of meaning and regulation.

PT REKI's success in accessing its ecosystem restoration concession in the first place indicates that the conservation company is also well-positioned with regard to the three power dimensions as well. Visible powers, especially the conservation company's material resources such as the ability to pay taxes in advance, were necessary to access the concession and establish the Harapan Rainforest project. Furthermore, PT REKI was able to change the rules of the game (hidden power). The conservation company's active lobbying led to the MoF reforming forest management, including introducing ecosystem restoration concessions that permit access to land in the first place. PT REKI also benefitted from its position with regard to the invisible dimension of power. First, PT REKI and partners were able to frame a former logging concession as one of the last remaining patches of "dry low land rainforest" and as one of the last remaining habitats for flagship species and for the indigenous Batin Sembilan. PT REKI strategically uses the presence of the Batin Sembilan and their alleged role as "forest-dependent people" to further legitimize forest conservation (Birdlife International, 2008: 5; Hein and Kunz, 2018: 160). Second, market-oriented conservation concepts such as private conservation concessions and REDD+ and the neoliberal consensus of privatization (Ellison, 2003; Harvey, 2005; Jenkins et al., 2004; McAfee, 1999, 2012a; Ong, 2006; Rodríguez de Francisco, 2013) laid the ground for PT REKI's successful lobbying activities. Moreover, the efforts of PT REKI and other conservation NGOs to construct a global scale of meaning to legitimize local conservation efforts has been internalized by some of the peasants interviewed in the study villages. Villagers, for instance, explained that "[…] the forest reserve is owned by the world"[2] and "[…] this forest is the lung of the earth",[3] accepting the designation of conservation areas for storing and capturing of greenhouse gases.

Territorialization

Territorialization is first of all a process of inclusion and exclusion of people. Territories are power relations written on land (Peluso and Lund, 2011) and a source of visible power. Territorialization as a process of claiming, naming and rule making is often considered as a precondition for accumulation. In Jambi, but also in other places, colonial and post-colonial state territorialization was an initial step in facilitating the exploitation of forests by private and state-owned companies. The formation of the Indonesian state forest territory separated people from land and other income sources and freed-up labor for the expanding agrobusiness. However, processes of primitive accumulation did not only occur in the context of the formation of the state forest, but they also occurred more recently in the context of the formation of protected areas. The Harapan Rainforest, for instance, restricts access to land but offers "green jobs" explicitly for the local population.

At Jambi's forest frontier, and also in other frontier areas, different territorialization processes compete, producing "contested spaces of sovereignty" (Agnew and Oslender, 2013: 124) characterized by often highly contested access and property relations. In Jambi, first of all, large amounts of forest are part of the state forest territory. The MoF allocated chunks of the state forest to timber companies and then recently to conservation companies. Still, parts of the state forest are simultaneously claimed by local and indigenous communities, village and district governments, and peasant organizations. Local and indigenous groups claim that these areas are part of their customary or ethnic territory. Many village governments expanded their village territories towards the state forest and thus facilitated the expansion of smallholder oil palm plantations. The peasant organization SPI considers its territory within the Harapan Rainforest as an alternative territory where land allocation is based on the social function of land outlined in the Basic Agrarian Law. The rules for land access developed by the different public authorities for these territories, as described in Chapter 3, are fundamentally different from those developed by the Ministry of Forestry for the state forest territory.

The nexus between property, authority and legitimacy

Access to natural resources including land is shaped by socio-spatial relations (Swyngedouw, 2010). But non-spatial social relations are also relevant for understanding the nexus between property, authority and legitimacy. Property, as outlined in the conceptual chapter, is a contested concept. The different meanings of property encountered in the landscapes of the Berbak Carbon Initiative and the Harapan Rainforest illustrate how changing ideologies and in particular a Western understanding of property have transformed property rights over time. Lineage-based property has been gradually replaced by individual forms of property that facilitate market exchange. Recently, new intangible types of property such as forest carbon have been introduced.

Conflictive or overlapping property rights can also be explained by unraveling conflicts over the "organizing ideology" of a society (Alagappa, 1995: 18; von Benda-Beckmann and von Benda-Beckmann, 1999: 30). Organizing ideology can be defined as the shared norms and truths of a society. In the case-study regions and in other parts of Indonesia, the organizing ideology is contested. In particular, *adat* ideology and different interpretations of Indonesia's state ideology clash in rural Jambi. Actors seek to legitimize property rights by referring to different elements of the competing ideologies. The social practices of Batin Sembilan, for instance, have been, and to a lesser extent still are, structured by *adat* ideologies. K. Benda-Beckmann and F. von Benda-Beckmann (1999: 30) argue that in many *adat* ideologies property rights over land should support and balance the livelihood of a community across generations. These elements are also reflected in the norms and beliefs of the Batin Sembilan. The property rights of the Batin Sembilan were attached to lineages and sub-lineages, providing access for all lineage members. To legitimize recent attempts to reestablish their former customary land (*wilayah adat*) the Batin Sembilan refer to *adat* and to powerful ancestors of the different lineages. In contrast, members of the peasant movement SPI, various apparatuses of the state (e.g. MoF, district and village governments) and the conservation company PT REKI base their arguments for legitimizing land claims on different interpretations of Indonesia's state ideology and on related laws and discourses. The peasant movement stresses the social function of land stated in the Basic Agrarian Law. The MoF stresses rather state ownership of forest land and the importance of forests to promote economic growth. The district head of Muaro Jambi highlighted the importance of Javanese "model farmers" promoting rural development in order to legitimize the formation of a transmigration settlement within the Berbak Carbon Initiative. He relates the allocation of land to nation building and to a development model dominated by a belief in the superiority of Javanese land-use practices.

The conservation company PT REKI refers to market environmentalism, in particular to the belief that state-based conservation approaches have failed. Therefore, it is argued, conservation needs to be privatized to maintain the carbon and biodiversity value of the Harapan Rainforest. The different legitimation strategies and entangled ideologies illustrate different meanings and functions of property: customary property as a source of community wealth, property as a means to promote socially inclusive rural development, property for promoting growth, and property as a way to protect ecosystems and trade ecosystem services.

In spite of the contested organizing ideology – or indeed because of it – the state or reference to the state is an important source of legitimacy, as Lund (2006: 690) has shown for Ghana and Niger. This is also the case in the study villages and in other parts of Sumatra (Kunz et al., 2016). To legitimize land claims as property, many actors refer to apparatuses of the state, symbols of the state, and language used by the state including laws, regulations and policies formulated by the state or its apparatuses. National regulation structures local agency. Local elites pick and choose certain state regulations that support their interests, e.g. the Basic Agrarian Law or Governmental Regulation 24/1997. The relevance of a certain land tenure regulation depends on the power structure in the arena. For example, in the area occupied by SPI, regulations formulated by the peasant movement are more relevant than the state regulations of the MoF.

Property is the enforceable right to objects of value. In consequence, property can only be considered as such if a legitimate public authority is able to sanction it. An important source of legitimacy for public authorities is again the state. At the village scale, the state is represented by a number of actors: by the elected village head, by hamlet heads, neighborhood heads and other members of the village government (*apparat desa*). Other important sources of legitimacy are state regulations, and also customary law, social identity and kinship. As Lund has shown for Ghana (Lund, 2008: 8), boundaries between the state and non-state actors are blurred. The village head in Bungku, for instance, claims to represent the state and to be a customary leader since he has kinship ties to a former lineage leader of the Batin Sembilan. The conservation company PT REKI enforces the Forest Law, has formulated additional conservation regulation, and issues de-facto land titles; the company has thus taken on a number of state functions. The legitimacy of property depends on acceptance of the authority issuing property rights. The acceptance of village-scale land titles across the study villages and the acceptance of village-scale land titles as collateral for accessing bank loans and as documents for accessing national-scale de-jure land titles indicate the legitimacy of the authority of village heads over land, irrespective of opposing laws and regulations.

Finally, an important source of legitimacy for public authorities is scale (Lund, 2006: 693). A number of different actors in both landscapes made reference to more powerful authorities at higher scales to legitimize property or their authority to issue property rights. In some cases, this again refers to the state and its various apparatuses, laws and regulations. In other cases, reference was made to "higher" customary authorities and even to international organizations and international law.

Networks of conservationists and of peasant resistance

Network relations also proved especially relevant for explaining land conflicts in the context of REDD+. Conservationists and peasants rely on transnational networks, as illustrated by the actor mappings in Chapter 5. Conservationists in particular have a distinct legacy of network building (Brockington et al., 2008: 9). Conservation projects often rely on private support and consequently often build on powerful alliances between corporate actors and conservationists (ibid. 7). The invention of the REDD+ mechanisms can be at least partly explained by such an alliance. One of the leading forest carbon standards, CCBS, for instance, has been supported from the very beginning by large companies such as BP and Intel. The Harapan Rainforest project as such was formed by an NGO network with three core members, Burung Indonesia, the Royal Society of Birds and Birdlife International, and other supporting members such as the German NABU (Naturschutzbund Deutschland), and has received support from a number of private companies. The network of actors involved in the implementation of the Berbak Carbon Initiative consists of NGOs, e.g. ZSL, Gita Buana, and state actors such as the Berbak National Park Agency and the Provincial Forest Agency of Jambi.

However, not only conservationists rely on transnational support networks. Peasants and indigenous groups have also received substantial support from transnational actors. For instance, La Via Campesina facilitated scale jumping for SPI, the NGOs CAPPA and SETARA support one another's lobbying activities, and alliances of village governments and customary leaders have raised the legitimacy of forest conversion activities. Private foundations, such as the Clinton and Ford Foundations, have also provided funding for advocacy groups which supported peasants and indigenous groups in conflicts with PT REKI.

Both conservationists and networks of peasants have close connections to state actors, again indicating the key role of the state in agrarian conflicts and in the privatization of conservation. As mentioned previously, only close connections to the Ministry of Forestry provided the opportunity for PT REKI and partners and also for ZSL to establish the conservation initiatives in the first place. In addition, both conservation networks relied on the support of public donors. Peasants mainly relied on alliances with village governments and specific sectoral state agencies.

Summing up

I have argued throughout this book that an explicit conceptualization of socio-spatial relations and of power and property is beneficial for political ecology. The different elements of a political ecology of transnational agrarian conflicts

I outlined above provide different entry points and perspectives into this rather new empirical phenomenon. A socio-spatial perspective helps to unravel the historical and spatial aspects of new rural enclosures such as conservation concessions, REDD+ projects and new agro-industrial estates. Different land-use practices and modes of production produce a specific spatial configuration which is often deeply entangled with access and property relations. A socio-spatial perspective helps to identify the specific historical geographies of social relations of conflict and cooperation that characterize transnational agrarian conflicts.

Investigating rescaling as one specific socio-spatial process helps to demarcate the arenas of socio-political struggle and regulation. Actors might produce additional scales or shift political struggles to a specific scale of regulation or political forum to pursue their interests (Hein et al., 2018a; Smith, 2008: 232; Towers, 2000; von Benda-Beckmann, 1981). New political scales are always the outcome of social practices. A new transnational scale of forest governance is a precondition for carbon trade and result-based payments and other forms of carbon finance. This new scale is currently under construction by actors interested in homogenous rules for carbon finance but also by indigenous groups seeking for international recognition. New protected areas are a specific type of territory and at the same time they introduce a new sub-national conservation scale, often controlled by NGOs and public conservation agencies, which also constitutes a new arena where social conflict between conservationists and opponents takes place.

A socio-spatial perspective considers dialectical relationships between structure and agency, an understanding of which is highly relevant. A spatial expansion of protest in the context of REDD+ attracted attention to structural inequality at the forest frontier, thus providing additional agency for indigenous communities. However, new green enclosures have also reduced agency significantly. An actor-centric approach, for example, is unlikely to grasp that socio-spatial structures impose significant (but changeable) limits on agency and resistance.

An explicit conceptualization of power helps reveal why socio-spatial relations may change. A three-dimensional view of power provides the means to uncover the conditions under which peasant resistance takes place in a hidden or open way. Rescaling and interrelated shifts in power relations, for instance in the context of decentralization, can explain why peasant resistance fails or succeeds.

Finally, I argue that different socio-spatial and non-spatial relations of transnational agrarian conflicts are characterized by at least three different dialectical relationships:

- Between scalar structure and agency
- Between power and scalar structure
- Between de jure and de-facto land and forest tenure regulations and their spatial dimensions.

Final remarks: implications for REDD+, uneven development and future directions of research for political ecology

This book has discussed the complexity of agrarian conflicts in the context of REDD+. But what does all this mean for the mechanism as such? First, clear land tenure is often mentioned as a precondition for the successful implementation of the mechanisms by scholars and practitioners (Galudra et al., 2014; Larson et al., 2013; Naughton-Treves and Wendland, 2014; Resosudarmo et al., 2014). Land is at the heart of the agrarian conflicts described in this book. But what exactly is meant by "clear" land and forest tenure? Indonesia's national REDD+ strategy refers to land tenure reform and to the "[. . .] constitutional rights to certainty over boundaries and management rights for natural resources" (Indonesian REDD+ Task Force, 2012: 18). At the village-scale, clear tenure might have different meaning than at the district or national scale. A land tenure reform would formalize certain customary rights but certainly not those of all right holders. Consequently, land tenure reform creates winners and losers. Second, a major success of the indigenous rights movement is their strong involvement in global debates on REDD+. Transnational safeguards for REDD+ projects acknowledge the rights of indigenous people and local communities, but the rights of non-local ethnic groups are frequently ignored. Conservation and REDD+ initiatives, such as the Harapan Rainforest, that judge the land rights of migrants as less legitimate may foster ethnic tensions. Third, it is argued that clear rights are relevant for selling forest carbon credits since the seller of carbon credits has to guarantee that the forest cover will remain for an agreed period (e.g. 30 years). If other actors claim the same forest area for other purposes, this could undermine the permanence of the avoided greenhouse gas emissions. Ongoing land conflicts about access to and control of the Harapan Rainforest and the Berbak Carbon Initiative as well as experiences from other countries implementing REDD+, such as Peru (Zelli et al., 2014), indicate that forest carbon offsetting is a very risky strategy to mitigate climate change. Instead of avoiding emissions, forest carbon offsetting could lead to additional greenhouse gas emissions if conflictive access and property relations undermine the integrity of forest areas designated for carbon offsetting.

Debates on the implementation of REDD+ as part of the Paris climate agreement are linked to questions of rural development and climate justice. REDD+ transfers part of the responsibility for reducing greenhouse gas emissions to rural areas of the Global South. The expansion of forest conservation may limit the development opportunities of those actors who have emitted the least fossil-fuel-based greenhouse gas emissions (e.g. see Irfany and Klasen, 2015 for emissions at household level in Indonesia). REDD+ and forest carbon offsetting are considered cheap mitigation options. However, the main reason why they are considered cheap are the lower opportunity costs of climate protection measures (such as REDD+) in the Global South (Hein, 2014: 510; McAfee, 2012b: 30). Thus, the success of REDD+ as an idea is based on uneven development and could even contribute to the persistence of uneven development.

Many political ecologists have conducted studies on the vulnerability of local populations in the context of climate change, climate variability and environmental change (Binternagel, 2011; Bohle et al., 1994; Bohle, 2011; Cutter et al., 2003; Few, 2003). However, the number of studies focusing on the impacts of climate mitigation policies on rural development and on the vulnerability of local populations is still limited and a rather new, and so far neglected, field of research (Cannon and Müller-Mahn, 2010; Horstmann and Hein, 2017). REDD+ and other climate protection policies should not increase the vulnerability to external shocks of the worst-off members of society. Many climate adaptation and mitigation measures are built on the assumption that development, e.g. transformation to a low-carbon economy, is a dirigible technical process and one that creates benefits for all actors. It is assumed that policy interventions such as investments in REDD+ lead to an anticipated outcome, e.g. to reduced deforestation rates, often omitting the fact that any policy shift creates winners and losers. The critical engagement of human geographers and political ecologists with climate protection instruments and with scalar dimensions of global climate governance could open up a relevant new frontier for geographical research. Political ecology could help to unravel the interests of actors investing in climate protection, power asymmetries between actors, and the legitimacy of private actors taking over former state functions. This could help support the design of policies that reduce greenhouse gas emissions and reduce the vulnerability of the worst-off members of society.

Notes

1 Today smallholders account for approximately 37 percent of Indonesia's annual oil palm production (Schwarze et al., 2015: 1)
2 Interview with a key informant in Kampung Laut, 29.08.2013, Document ID: 135.
3 Interview with a key informant in Bungku, 08.09.2012, Document ID: 77.

References

Acciaioli, G., 2001. Grounds of conflict, idioms of harmony: Custom, religion, and nationalism in violence avoidance at the Lindu Plain, Central Sulawesi. *Indonesia*, (72), 81–114.

Adiwibowo, S., 2005. *Dongi-Dongi-Culmination of a multi-dimensional ecological crisis: A political ecology perspective.* Thesis (PhD). University Kassel, Germany.

Agnew, J., and U. Oslender. 2013. Overlapping territorialities, sovereignty in dispute: Empirical lessons from Latin America. *In:* W. Nicholls, B. Miller and J. Beaumont, eds. *Spaces of Contention: Spatialities and Social Movements.* Barlington: Ashgate, 191–213.

AGRA. 2015. *Aliansi Gerakan Reforma Agraria (AGRA) "Tidak Ada Demokasi Tanpa Land Reform".* Available from: https://agraindonesia.org/profil/ [Accessed 06 June 2018].

AGRA. 2017. *Siaran Pers AGRA Cabang Jambi Terkait Bantahan PT. REKI Atas Pembakaran Rumah Suku Anak Dalam.* Available from: http://agraindonesia.org/press-release-aliansi-gerakan-reforma-agraria-agra-cabang-jambi-terkait-bantahan-pt-reki-terhadap-sikap-agra-atas-pembakaran-rumah-sad-oleh-pt-reki/ [Accessed 21 August 2018].

Alagappa, M., 1995. *Political Legitimacy in Southeast Asia: The Quest for Moral Authority.* Stanford, CA: Stanford University Press.

Alamsyah, Z., 2004. *Socio-Economic Conditions in Communities in the Vicinity of Berbak National Park, A Rapid Appraisal in Air Hitam Laut Village and Sungai Gelam Village.* Wageningen: The Netherlands Water for Food and Ecosystems Programme, Wageningen University.

Allen, J., 2003. Power. *In:* J. Agnew and G. Toal, eds. *A Companion to Political Geography.* Malden, MA: Blackwell Publishing Ltd, 95–108.

Aliansi Masyarakat Adat Nusantara (AMAN). 2015. *Profil Organisasi.* Available from: www.aman.or.id/wp-content/uploads/2015/11/Profile-AMAN_2015.pdf [Accessed 27 November 2015].

Aliansi Masyarakat Adat Nusantara (AMAN) Bengkulu. 2014. *AMAN dan SPI Klaim Sumbang 33 Ribu Suara.* Available from: http://amanbengkulu.or.id/aman-dan-spi-klaim-sumbang-33-ribu-suara/ [Accessed 15 November 2015].

Andaya, B. W., 1993. *To live as brothers: Southeast Sumatra in the seventeenth and eighteenth centuries.* Honolulu: University of Hawaii Press.

Andaya, L. Y., 2008. *Leaves of the Same Tree: Ethnicity and trade in the Straits of Melaka.* Honolulu. Honolulu: The University of Hawai'i Press.

Angelsen, A., M. Brockhaus, W. D. Sunderlin, and L. Verchot. 2012. Introduction. *In:* A. Angelsen, M. Brockhaus, W. D. Sunderlin and L. Verchot, eds. *Analysing REDD+: Challenges and Choices.* Bogor, Indonesia: Center for International Forestry Research (CIFOR), 1–12.

Angelsen, A., C. W. Gierløff, A. M. Beltrán, and M. den Elzen. 2014. *REDD Credits in a Global Carbon Market: Options and Impacts.* Copenhagen: Nordic Council of Ministers.

Armitage, D., 2002. Socio-institutional dynamics and the political ecology of mangrove forest conservation in Central Sulawesi, Indonesia. *Global Environmental Change*, 12 (3), 203–217.

Aspinall, E., 2013. A nation in fragments: Patronage and neoliberalism in contemporary Indonesia. *Critical Asian Studies*, 45(1), 27–54

Bachriadi, D., and G. Wiradi. 2011. *Six Decades of Inequality*. Bandung, Indonesia: Agrarian Resource Centre, Bina Desa, Konsortium Pembaruan Agraria.

Bäckstrand, K., and E. Lövbrand. 2006. Planting trees to mitigate climate change: Contested discourses of ecological modernization, green governmentality and civic environmentalism. *Global Environmental Politics*, 6 (1), 50–75.

Badan Inventarisasi dan Tata Guna Hutan. n.a. Peta Areal Hutan Yang Disetujui untuk Dilepaskan Guna Peruntukan Perkebunan An. PT Bangun Desa Utama Jakarta, Indonesia: Departemen Kehutanan.

Badan Inventarisasi dan Tata Guna Hutan. 1985. *Peta Rencana Pengukuhan dan Penatagunaan Hutan, Propinsi Dati I Jambi*. Jakarta: Indonesia Departemen Kehutanan.

Badan Inventarisasi dan Tata Guna Hutan. 1993. *Rencana Struktur Tata Ruang Propinsi, Propinsi Jambi*. Jakarta, Indonesia: Departemen Kehutanan.

Badan Pengelola REDD+. 2014. *Kegiatan DA REDD+ di Jambi*. Available from: http://kc.reddplusid.org/kegiatan-dan-program/provinsi-prioritas/jambi/480-kegiatan-da- [Accessed 11 February 2015].

Bakhori, S., 2013. *853.430 Hektare Hutan di Jambi Dikuasai HTI*. Available from: https://m.tempo.co/read/news/2013/08/16/058504870/853-430-hektare-hutan-di-jambi-dikuasai-hti [Accessed 29 May 2016].

Bakker, L., 2008. "Can We Get Hak Ulayat?": Land and community in pasir and nunukan, East Kalimantan. In: *UC Berkeley-UCLA Joint Conference on Southeast Asia, "Ten Years After: Reformasi and New Social Movements in Indonesia, 1998–2008*. Berkeley, CA: Center for Southeast Asia Studies, UC Berkeley.

Bakker, L., and S. Moniaga. 2010. The space between: Land claims and the law in Indonesia. *Asian Journal of Social Science*, 38 (2), 187–203.

Balai Taman Nasional Berbak. 2013. *Sejarah Pembentukan Taman Nasional Berbak*. Jambi, Indonesia.

Balai Taman Nasional Berbak and Zoological Society London. 2011. Perjanjian Kerjasama antara Balai Taman Nasional Berbak (BTNB) and the Zoological Society London (ZSL). *SP 42//BTNB-1/2011*. Jambi, Indonesia.

Balan, S., 2010. M. Foucault's view on power relations. *Revista Cogito*, 2 (2), 1–6.

Barr, C., 2006. Forest administration and forestry sector development prior to 1998. *In*: C. Barr, I. A. P. Resosudarmo, A. Dermawan and J. McCarthy, eds. *Decentralization of Forest Administration in Indonesia: Implications for Forest Sustainability, Economic Development and Community Livelihoods*. Bogor, Indonesia: Center for international Forestry Research (CIFOR), 18–30.

Barr, C. M., 1998. Bob Hasan, the rise of APKINDO, and the shifting dynamics of control in Indonesia's timber sector. *Indonesia*, (65), 1–36.

Barr, C., I. Resosudarmo, J. McCarthy, and A. Dermawan. 2006. Forests and decentralization in Indonesia: An overview. *In*: C. R. Barr, I. A. P. Resosudarmo, A. Dermawan and J. McCarthy, eds. *Decentralization of Forest Administration: Implications for Forest Sustainability, Economic Development and Community Livelihoods*. Bogor, Indonesia: Center for international Forestry Research (CIFOR), 1–17.

Bebbington, A., L. Dharmawan, E. Fahmi, and S. Guggenheim. 2004. Village politics, culture and community-driven development: Insights from Indonesia. *Progress in Development Studies*, 4 (3), 187–205.

Beckert, B., C. Dittrich, and S. Adiwibowo. 2014. Contested land: An analysis of multi-layered conflicts in Jambi province, Sumatra, Indonesia. *Austrian Journal of South-East Asian Studies*, 7 (1), 75–92.

Beckert, B., and M. Keck. 2015. Palmöl für den Weltmarkt: Landkonflikte in Sumatras Post-Frontier. *Geographische Rundschau* 67 (12), 12–17.

Beckert, B. 2017. A post-frontier in transformation: land relations between access, exclusion and resistance in Jambi province, Indonesia. PhD-thesis, Department of Human Geography, Niedersächsische Staats-und Universitätsbibliothek Göttingen, Göttingen, Germany.

Bedner, A., and S. Van Huis. 2008. The return of the native in Indonesian law: Indigenous communities in Indonesian legislation. *Bijdragen tot de Taal-, Land-en Volkenkunde/Journal of the Humanities and Social Sciences of Southeast Asia*, 164 (2–3), 165–193.

Behrens, M., and H. Janusch. 2012. Der transnationale Wettbewerbsstaat. *Journal für Entwicklungspolitik*, 28(2), 28–53.

Benoit, D., and O. Sevin. 1993. L'émigration javanaise: Mythes et réalités. *Annales de géographie*, 102 (571), 255–276.

Bhan, M., D. Sharma, A. Ashwin, and S. Mehra. 2017. Policy forum: Nationally-determined climate commitments of the BRICS: At the forefront of forestry-based climate change mitigation. *Forest Policy and Economics*, 85, 172–175.

Binternagel, N., 2011. Adaptation to natural hazards in Central Sulawesi, Indonesia-strategies of rural households. PhD-thesis, Department of Human Geography, Niedersächsische Staats-und Universitätsbibliothek Göttingen, Göttingen, Germany.

Birdlife International. 2008. *Long Term Conservation of the Harapan Rainforest in Sumatra, Final Report to The Nando Peretti Foundation*. Cambridge, UK and Jakarta, Indonesia: Birdlife International.

Birdlife International. 2015. *About Birdlife*. Available from: www.birdlife.org/worldwide/partnership/about-birdlife [Accessed 04 December 2015].

Blaikie, P., 2012. Should some political ecology be useful? The Inaugural Lecture for the Cultural and Political Ecology Specialty Group, Annual Meeting of the Association of American Geographers, April 2010. *Geoforum*, 43 (2), 231–239.

Blaikie, P., and H. Brookfield. 1987. *Land Degradation and Society*. London, UK: Methuen.

Bohle, H. G., 2011. Geographische Entwicklungsforschung *In*: H. Gebhardt, R. Glaser, U. Radtke and P. Reuber, eds. *Geographie, Physissche Geographie und Humangeographie*. München: Spektrum Akademischer Verlag, 745–763, 779–783.

Bohle, H. G., T. E. Downing, and M. J. Watts. 1994. Climate change and social vulnerability: Toward a sociology and geography of food insecurity. *Global Environmental Change*, 4 (1), 37–48.

Bohle, H. G., and H. Fünfgeld. 2007. The political ecology of violence in eastern Sri Lanka. *Development and Change*, 38 (4), 665–687.

Borras, Jr, S. M., 2008. La Vía Campesina and its global campaign for agrarian reform. *Journal of Agrarian Change*, 8 (2–3), 258.

Brad, A., 2016. Politische Ökologie und Politics of Scale-Vermittlungszusammenhänge zwischen Raum, Natur und Gesellschaft. *Geographica Helvetica*, 71 (4), 353.

Brad, A., A. Schaffartzik, M. Pichler, and C. Plank. 2015. Contested territorialization and biophysical expansion of oil palm plantations in Indonesia. *Geoforum*, 64, 100–111.

Brand, U., and C. Görg. 2003. The state and the regulation of biodiversity: International biopolitics and the case of Mexico. *Geoforum*, 34 (2), 221–233.

Brand, U., C. Görg, J. Hirsch, and M. Wissen. 2008. The regulation of nature in post-Fordism. *In*: U. Brand, C. Görg, J. Hirsch and M. Wissen, eds. *Conflicts in Environmental Regulation and the Internationalisation of the State*. Milton Park: Routledge, 9–52.

Brand, U., C. Görg, and M. Wissen. 2011. Second-order condensations of societal power relations: Environmental politics and the internationalization of the state from a neo-poulantzian perspective. *Antipode*, 43 (1), 149–175.

Brand, U., and M. Wissen. 2012. Global environmental politics and the imperial mode of living: articulations of state – capital relations in the multiple crisis. *Globalizations*, 9, 4.

Brand, U., and M. Wissen. 2017. *Imperiale Lebensweise: Zur Ausbeutung von Mensch und Natur im globalen Kapitalismus*. München: Oikom.

Brenner, N., 1997. Global, fragmented, hierarchical: Henri Lefebvre's geographies of globalization. *Public Culture*, 10 (1), 135–167.

Brenner, N., 1998. Between fixity and motion: Accumulation, territorial organization and the historical geography of spatial scales. *Environment and Planning D*, 16, 459–482.

Brenner, N., 2001. The limits to scale? Methodological reflections on scalar structuration. *Progress in Human Geography*, 25 (4), 591–614.

Brockhaus, M., K. Obidzinski, A. Dermawan, Y. Laumonier, and C. Luttrell. 2012. An overview of forest and land allocation policies in Indonesia: Is the current framework sufficient to meet the needs of REDD+? *Forest Policy and Economics*, 18, 30–37.

Brockington, D., R. Duffy, and J. Igoe. 2008. *Nature Unbound: Conservation, Capitalism, and the Future of Protected Areas*. London, UK: Earthscan Publishers.

Brockington, D., and L. Scholfield. 2011. The conservationist mode of production and conservation NGOs in sub-Saharan Africa. *In*: D. Brockington and R. Duffy, eds. *Capitalism and Conservation*. West Sussex, UK: Wiley-Blackwell, 82–107.

Bryant, R. L., 1998. Power, knowledge and political ecology in the third world: A review. *Progress in Physical Geography*, 22 (1), 79–94.

Bryant, R. L., 2001. Political ecology: A critical agenda for change? *In*: N. Castree and B. Braun, eds. *Social Nature: Theory, Practice, and Politics*. Malden, MA and Oxford, UK: Blackwell Publishers Ltd, 151–169.

Buckley, K., 2018. Space, social relations, and contestation: Transformative peacebuilding and world social forum climate spaces. *Antipode*, 50 (2), 279–297.

Buergin, R., 2014. *Forest Problematic and Bilateral Forest Related German Development Cooperation in the Case Study Country Indonesia*. Freiburg im Breisgau, Germany: Institute of Forest Sciences, University of Freiburg.

Buletin Batin Sembilan. 2013. *A New Round in the Meeting between PT. REKI and the Communities of Simpang Macan Luar dan Bawah Bedaro*. Jambi: Yayasan Cappa, Setara, AGRA, Perkumpulan Hijau.

Bulkeley, H., 2005. Reconfiguring environmental governance: Towards a politics of scales and networks. *Political Geography*, 24 (8), 875–902.

Bulkeley, H., and P. Newell. 2015. *Governing Climate Change*. London, UK: Routledge.

Burkard, G., 2002. *Stability or sustainability? Dimensions of socio-economic security in a rain forest margin. Vol. No. 7 (September 2002), STORMA Discussion Paper Series*. Göttingen, Germany: STORMA Discussion Paper Series.

Cabello, J., and T. Gilbertson. 2012. A colonial mechanism to enclose lands: A critical review of two REDD+-focused special issues. *Ephemera*, 12 (1/2), 162.

Cannon, T., and D. Müller-Mahn. 2010. Vulnerability, resilience and development discourses in context of climate change. *Natural Hazards*, 55 (3), 621–635.

Caouette, D., and S. Turner. 2009a. Rural resistance and the art of domination. *In*: D. Caouette and S. Turner, eds. *Agrarian angst and rural resistance in contemporary Southeast Asia*, Abingdon, UK: Routledge, 25–44.

Caouette, D., and S. Turner. 2009b. Shifting fields of rural resistance in Southeast Asia. *In*: D. Caouette and S. Turner, eds. *Agrarian Angst and Rural Resistance in Contemporary Southeast Asia*. Abingdon, UK: Routledge, 1–24.

Carr, D., 2009. Population and deforestation: Why rural migration matters. *Progress in Human Geography*, 33 (3), 355–378.

Casson, A., and K. Obidzinski. 2002. From new order to regional autonomy: Shifting dynamics of "Illegal" logging in Kalimantan, Indonesia. *World Development*, 30 (12), 2133–2151.

Castillo, S., 2013. *How Costa Rica and Chile are Leveraging Independent Carbon Standards to Get Ready for REDD*. Available from: www.forest-trends.org/ecosystem_marketplace/how-costa-rica-and-chile-are-leveraging-independent-carbon-standards-to-get-ready-for-redd/ [Accessed 19 June 2018].

Castree, N., 2001. Socializing Nature: Theory, Practice, and Politics. *In*: N. Castree and B. Braun, eds. *Social Nature: Theory, Practice, and Politics*. Oxford, UK: Blackwell Publishers Ltd, 1–21.

Castree, N., 2003. Commodifying what nature? *Progress in Human Geography*, 27 (3), 273–297.

Castree, N., 2008. Neoliberalising nature: Processes, effects, and evaluations. *Environment and planning. A*, 40 (1), 153.

CCBA, Climate, Community & Biodiversity Alliance. 2008. *Climate, Community and Biodiversity Project Design Standards, Second Edition*. Arlington, VA: CCBA.

CCBA, Climate, Community & Biodiversity Alliance. 2015. *About the CCBA*. Available from: www.climate-standards.org/about-ccba/ [Accessed 04 December 2015].

CCBA, Climate, Community & Biodiversity Alliance. 2018. *History of the Standards*. Available from: http://www.climate-standards.org/ccb-standards/history-of-the-standards/ [Accessed 23 August 2018].

Chan, S., H. Asselt, T. Hale, K. W. Abbott, M. Beisheim, M. Hoffmann, B. Guy, N. Höhne, A. Hsu, P. Pattberg, P. Pauw, C. Ramstein, and O. Widerberg. 2015. Reinvigorating international climate policy: A comprehensive framework for effective nonstate action. *Global Policy*, 6 (4), 466–473.

Chan, S., C. Brandi, and S. Bauer. 2016. Aligning transnational climate action with international climate governance: The road from Paris. *Review of European, Comparative & International Environmental Law*, 25 (2), 238–247.

Chatterton, P., D. Featherstone, and P. Routledge., 2013. Articulating climate justice in Copenhagen: Antagonism, the commons, and solidarity. *Antipode*, 45 (3), 602–620.

Chin, C. B. N., and J. H. Mittelman. 1997. Conceptualising resistance to globalisation. *New Political Economy*, 2 (1), 25–37.

CIFOR, 2012. *REDD+ project sites in Indonesia*. Available from: www.forestsclimate change.org/global-comparative-study-on-redd/redd-project-sites/redd-project-sites-in-indonesia.html [Accessed 04 December 2015].

Ciplet, D., 2014. Contesting climate injustice: Transnational advocacy network struggles for rights in UN climate politics. *Global Environmental Politics*, 14 (4), 75–96.

CIRAD, French Agricultural Research Centre for International Development, 2012. *ZSL Indonesia Field Programme*. Available from: http://spop.cirad.fr/project/documents/project-benchmarking/zsl-field-programme [Accessed 04 December 2015].

Claeys, P., and D. Delgado Pugley. 2017. Peasant and indigenous transnational social movements engaging with climate justice. *Canadian Journal of Development Studies/Revue canadienne d'études du développement*, 38 (3), 325–340.

Claridge, G., 1994. Management of coastal ecosystems in eastern Sumatra: The case of Berbak Wildlife Reserve, Jambi Province. *Hydrobiologia*, 285 (1–3), 287–302.

Cohen, A., and J. McCarthy. 2014. Reviewing rescaling: Strengthening the case for environmental considerations. *Progress in Human Geography*, 39 (1), 3–25.

Colchester, M., P. Anderson, A. Y. Firdaus, F. Hasibuan, and S. Chao. 2011. *Human Rights Abuses and Land Conflicts in the PT Asiatic Persada Concession in Jambi: Report of an Independent Investigation into Land Disputes and Forced Evictions in a Palm Oil Estate: Independent Investigation of PT AP*. Moreton-in-Marsh, UK, Bogor and Jakarta, Indonesia: HuMa, Sawit Watch, Forest Peoples Programme.

Conservation International, Environmental Defense Fund, Forest Trends, IETA Climate Challenges Market Solutions and The Nature Conservancy. 2016. *Inputs from Conservation International, Environmental Defense Fund, Forest Trends, International Emissions Trading Association and The Nature Conservancy Regarding Views on the Guidance Referred to in Article 6, Paragraph 2 of the Paris Agreement*. Available from: https://www.conservation.org/publications/Docu ments/CI_UNFCCC-SB46-Submission1-Article6.2.pdf [Accessed 25 September 2018].

Corbera, E., 2012. Problematizing REDD+ as an experiment in payments for ecosystem services. *Current Opinion in Environmental Sustainability*, 4 (6), 612–619.

Corbera, E., and K. Brown. 2010. Offsetting benefits? Analyzing access to forest carbon. *Environment and planning. A*, 42 (7), 1739–1761.

Corbera, E., and H. Schroeder. 2017. REDD+ crossroads post Paris: Politics, lessons and interplays. *Forests*, 8 (12), 508.

Corbera, E., C. G. Soberanis, and K. Brown. 2009. Institutional dimensions of Payments for Ecosystem Services: An analysis of Mexico's carbon forestry programme. *Ecological Economics*, 68 (3), 743–761.

Cramb, R. A., C. J. P. Colfer, W. Dressler, P. Laungaramsri, Q. T. Le, E. Mulyoutami, N. L. Peluso, and R. L. Wadley. 2009. Swidden transformations and rural livelihoods in Southeast Asia. *Human Ecology*, 37 (3), 323–346.

Cutter, S. L., B. J. Boruff, and W. L. Shirley. 2003. Social vulnerability to environmental hazards. *Social Science Quarterly*, 84 (2), 242–261.

DANIDA. 2004. *Strategy for Danish Support to Indigenous Peoples*. Copenhagen: Ministry of Foreign Affairs.

DANIDA. 2011. *How to Note on Indigenous Peoples*. Available from: http://um.dk/en/~/media/UM/English-site/Documents/Danida/Activities/Strategic/Human%20rights%20and%20democracy/Human%20rights/How%20to%20Note%20Indigenous%20Peoples.pdf [Accessed 25 June 2018].

DANIDA. 2012a. *Harapan Rainforest Project in Indonesia*. Available from: http://um.dk/en/danida-en/activities/annual-report-2012/programmes-and-projects/green-growth-hara-pan-rainforest-project-in-indonesia/ [Accessed 24 July 2015].

DANIDA. 2012b. *The Right to a Better Life Strategy for Denmark's Development Cooperation*. Available from: http://um.dk/en/~/media/UM/Danish-site/Documents/Danida/Det-vil-vi/right_to_a_better_life_pixi.pdf [Accessed 21 August 2018].

DANIDA. 2015. *ESP3 DANIDA Environmental Support Programme, Annual Report*. Jakarta: DANIDA.

DANIDA. 2016. *ESP3 DANIDA Environmental Support Programme, Annual Progress Report*. Jakarta: DANIDA.

DANIDA. 2016. *Restricted Procedure: Financial Adviser to Burung Indonesia, Hutan Harapan Phase III – Indonesia*. Available form: http://um.dk/en/about-us/procurement/contracts/short/contract-opportunitie/newsdisplaypage/?newsid=927adc66-9b61-43c9-8393-cc59871050c1 [Accessed 01 June 2018].

Davidson, J. S., and D. Henley. 2007. Introduction: Radical conservatism – the protean politics of adat. *In*: J.S. Davidson and D. Henley, eds. *The Revival of Tradition in Indonesian Politics*. London, UK: Routledge, 21–69.

Del Cairo, C., I. Montenegro-Perini, and J. S.Vélez. 2014. Naturalezas, subjetividades y políticas ambientales en el Noroccidente amazónico: reflexiones metodológicas para el análisis de conflictos socioambientales. *Boletín de Antropología*, 29 (48), 13–40.

Delgado-Pugley, D., 2013. Contesting the limits of consultation in the Amazon Region: On indigenous peoples' demands for free, prior and informed consent in Bolivia and Peru. *Revue générale de droit*, 43, 151–181.

Demirović, A., 2011. Materialist state theory and the transnationalization of the capitalist state. *Antipode*, 43 (1), 38–59.

Dewan Nasional Perubahan Iklim. 2010. *Creating Low Carbon Prosperity in Jambi*. Jambi and Jakarta, Indonesia.

Díaz, C. A. B., 2014. Contested lands, contested laws. *Americas Quarterly*, 8 (2), 106.

Dinas Kehutanan Kabupaten Batang Hari. 2012. *Laporan Hasil Identifikasi dan Inventarisasi Masyarakat penggarap Wilayah Bukit Sinyak, Areal Kerja PT. Restorasi Ekosistem Indonesia*. Muaro Bulian, Jambi and Indonesia: Kabupaten Batang Hari.

Dinas Kehutanan Provinsi Jambi. 2013. *Persetujuan Lokasi Demonstration Activities Reducing Emission from Deforestation and Forest Degradation (DA-REDD+) di Taman Hutan Raya Tanjung, 81/A/ZSL/II/2013*. Jambi, Indonesia: Pemerintah Provinsi Jambi.

Dinas Komunikas dan Informatika Provinsi Jambi. 2013. *Sejarah Berdirinya Provinsi Jambi*. Available from: http://jambiprov.go.id/index.php?sejarah [Accessed 03 December 2015].

Dinas Tenaga Kerja dan Transmigrasi Sumatera Barat. 2015. *Bursa Transmigrasi*. Available from: http://bto-sumbar.com/index.php/en/ [Accessed 03 December 2015].

Direktorat Jenderal Perkebunan. 2017. *Statistik Perkebunan Indonesia, 2015–2017, Kelapa Sawit*. Jakarta: Sekretariat Direktorat Jenderal Perkebunan, Direktorat Jenderal Perkebunan and Kementerian Pertanian.

Duchelle, A. E., C. de Sassi, P. Jagger, M. Cromberg, A. M. Larson, W. D. Sunderlin, S. S. Atmadja, I. A. P. Resosudarmo, C. D. Pratama. 2017. Balancing carrots and sticks in REDD+ implications for social safeguards. *Ecology and Society*, 22 (3).

Duchelle, A. E., M. Greenleaf, D. Mello, M. F. Gebara, and T. Melo. 2014. Acre's State System of Incentives for Environmental Services (SISA), Brazil. *In*: E. O. Sills, S. S. Atmadja, C. de Sassi, A. E. Duchelle, D. L. Kweka, I. A. P. Resosudarmo and W. D. Sunderlin, eds. *REDD+ on the Ground*. Bogor, Indonesia: Center for international Forestry Research (CIFOR), 33–50.

Eickhoff, G., A. Salim, and S. A. Stanley. 2010. *Initial Field and Desktop Assessment of Carbon Emission Reduction Potential for the Berbak Carbon Initiative*. Jakarta, Indonesia: Forest Carbon.

Eilenberg, M., 2015. Shades of green and REDD: Local and global contestations over the value of forest versus plantation development on the Indonesian forest frontier. *Asia Pacific Viewpoint*, 56 (1), 48–61.

Ekers, M., and A. Loftus. 2008. The power of water: Developing dialogues between Foucault and Gramsci. *Environment and Planning D: Society and Space*, 26 (4), 698–718.

Ekers, M., A. Loftus, and G. Mann. 2009. Gramsci lives! *Geoforum*, 40 (3), 287–291.

Eliasch, J., 2008. *The Eliasch Review – Climate Change: Financing Global Forests. Commissioned by The Office of Climate Change, UK*. Richmond, UK: The Stationery Office Limited on behalf of the Controller of Her Majesty's Stationery Office.

Ellison, K., 2003. Renting biodiversity: The conservation concessions approach. *Conservation in Practice*, 4 (4), 20–29.

Elmhirst, R., 2001. Resource struggles and the politics of place in North Lampung, Indonesia. *Singapore Journal of Tropical Geography*, 22 (3), 284–306.

England, K.V., 1994. Getting personal: Reflexivity, positionality, and feminist research★. *The Professional Geographer*, 46 (1), 80–89.

Escobar, A., 1996. Construction nature: Elements for a post-structuralist political ecology. *Futures*, 28 (4), 325–343.

Escobar, A., 1999. After nature: Steps to an antiessentialist political ecology 1. *Current Anthropology*, 40 (1), 1–30.

Escobar, A., 2003. Displacement, development, and modernity in the Colombian Pacific. *International Social Science Journal*, 55 (175), 157–167.

Etzold, B., M. Keck, H. G. Bohle, and W. P. Zingel. 2009. Informality as agency – negotiating food security in Dhaka. *Die Erde*, 140 (1), 3–24.

Fairhead, J., M. Leach, and I. Scoones. 2012. Green grabbing: A new appropriation of nature? *Journal of Peasant Studies*, 39 (2), 237–261.

FAO. 2014. *State of the World's Forests – Enhancing the Socioeconomic Benefits from Forests*. Rome: FAO.

Fauna & Flora International. 2012. *Community Forest Ecosystem Services, Indonesia Plan Vivo Project Idea Note (PIN)*. Jakarta, Indonesia: FFI Indonesia Programme.

Faust, H., 2007. *Vergleichende Kulturgeographie: Empirische Befunde regionaler Integrationsprozesse in tropischen Agrarkolonisationsräumen Boliviens, der Elfenbeinküste und Indonesiens*. Göttingen, Germany: Goltze.

Faust, H., S. Schwarze, B. Beckert, B. Brümmer, C. Dittrich, M. Euler, M. Gatto, B. Hauser-Schäublin, J. Hein, and A. Holtkamp. 2013. *Assessment of socio-economic functions of tropical lowland transformation systems in Indonesia*. Vol. No. 1, *EFForts Discussion Paper Series*. Göttingen, Germany: University of Göttingen.

Fearnside, P. M., 1997. Transmigration in Indonesia: Lessons from its environmental and social impacts. *Environmental Management*, 21 (4), 553–570.

Ferdiyal, I., 2013. *Pengusiran Petani REKI, Komnas HAM Temukan Indikasi Pelanggaran HAM*. Available from: www.metrojambi.com/v1/daerah/13893-pengusiran-petani-reki-komnas-ham-temukan-indikasi-pelanggaran-ham.html [Accessed 15 September 2015].

Few, R., 2003. Flooding, vulnerability and coping strategies: Local responses to a global threat. *Progress in Development Studies*, 3 (1), 43–58.

Finlayson, R., 2014. *REDD Ready or Not? Agroforestry World Blog*. Available from: http://blog.worldagroforestry.org/index.php/2014/06/19/redd-ready-or-not/ [Accessed 11 November 2015].

Fisher, J., 2012. No pay, no care? A case study exploring motivations for participation in payments for ecosystem services in Uganda. *Oryx*, 46 (1), 45–54.

Fletcher, R., 2010. Neoliberal environmentality: Towards a poststructuralist political ecology of the conservation debate. *Conservation and Society*, 8 (3), 171.

Flitner, M., and C. Görg. 2008. Politik im Globalen Wandel – räumliche Maßstäbe und Knoten der Macht. *In*: A. Brunnengräber, H. J. Burchardt and C. Görg, eds. *Mit mehr Ebenen zu mehr Gestaltung? Multi-Level-Governance in der transnationalen Sozial- und Umweltpolitik*. Baden-Baden, Germany: Nomos Verlag, 163–181.

Fogel, C., 2004. The local, the global, and the Kyoto Protocol. *In*: S. Jasanoff and M. L. Martello, eds. *Earthly Politics: Local and Global in Environmental Governance*. Cambridge, MA: MIT Press, 103–125.

Fold, N., and P. Hirsch. 2009. Re-thinking frontiers in Southeast Asia. *The Geographical Journal*, 175 (2), 95–97.

Forest Climate Center. 2012. *REDD Project List – Indonesia (March 2012 Version)*. Available from: http://forestclimatecenter.org/files/2012-03-26%20Indonesia%20-%20REDD%20Demonstration%20Activities.pdf [Accessed 04 December 2015].

Forest Trends. 2014. *REDDX Tracking Forest Finance, Indonesia*. Available from: http://reddx.forest-trends.org/country/indonesia/overview [Accessed 04 December 2015].

Forsyth, T., 2008. Political ecology and the epistemology of social justice. *Geoforum*, 39 (2), 756–764.

Foucault, M., 2006. *Sicherheit, Territorium, Bevölkerung: Geschichte der Gouvernementalität I. Vorlesungen am Collège de France 1977–1978, Geschichte der Governementalität.* Frankfurt am Main, Germany: Suhrkamp.

Galudra, G., and M. Sirait. 2009. A discourse on Dutch colonial forest policy and science in Indonesia at the beginning of the 20th century. *International Forestry Review*, 11 (4), 524–533.

Galudra, G., M. van Noordwijk, P. Agung, S. Suyanto, and U. Pradhan. 2014. Migrants, land markets and carbon emissions in Jambi, Indonesia: Land tenure change and the prospect of emission reduction. *Mitigation and Adaptation Strategies for Global Change*, 19 (6), 715–731.

Galudra, G., M.Van Noordwijk, S. Suyanto, I. Sardi, U. Pradhan, and D. Catacutan. 2011. Hot spots of confusion: Contested policies and competing carbon claims in the peatlands of Central Kalimantan, Indonesia. *International Forestry Review*, 13 (4), 431–441.

Gamin, B. N., H. Kartodihardjo, L. M. Kolopaking, and R. Boer. 2014. Forest land tenure conflicts and their resolutions: A case study from Musi Rawas Regency, South Sumatera Province. *Journal of Islamic Perspective on Science, Technology and Society*, 2 (1), 53–64.

Garland, E., 2008. The elephant in the room: Confronting the colonial character of wildlife conservation in Africa. *African Studies Review*, 51 (3), 51–74.

Gaventa, J., 1982. *Power and powerlessness: Quiescence and Rebellion in an Appalachian Valley.* Champaign: University of Illinois Press.

Gaventa, J., 2003. *Power after Lukes: An Overview of Theories of Power Since Lukes and their Application to Development.* Brighton, UK: Institute of Development Studies.

Gaventa, J., 2006. Finding the spaces for change: A power analysis. *IDS Bulletin*, 37 (6), 23–33.

Gaventa, J., J. Pettit, and L. Cornish. 2011. *Scott: Resistance.* Available from: www.power cube.net/other-forms-of-power/scott-resistance/ [Accessed 14 October 2015].

Giesen, W., 2004. *Causes of Peat Swamp Forest Degradation in Berbak NP, Indonesia, and Recommendations for Restoration, Water for Food and Ecosystems Programme Project on: "Promoting the River Basin and Ecosystem Approach for Sustainable Management of SE Asian Lowland Peat Swamp Forests: Case Study Air Hitam Laut River basin, Jambi Province, Sumatra, Indonesia".* Arnheim, The Netherlands: ARCADIS Euroconsult.

Gita Buana., 2013. *Progres Kegiatan FGD-FPIC di Kecamatan Kumpeh, Kabupaten Muaro Jambi* (Unpublished document). Jambi, Indonesia.

Gómez, C. J., L. Sánchez-Ayala, and G. A. Vargas. 2015. Armed conflict, land grabs and primitive accumulation in Colombia: micro processes, macro trends and the puzzles in between. *Journal of Peasant Studies*, 42(2): 255–274.

Görg, C., 1999. *Gesellschaftliche Naturverhältnisse.* Münster: Westfälisches Dampfboot.

Görg, C., 2011. Shaping relationships with nature – adaptation to climate change as a challenge for society. *Die Erde*, 142 (4), 411–428.

Government of the Kingdom of Norway and Government of the Republic of Indonesia. 2010. *Letter of Intent Between the Government of the Kingdom of Norway and Government of the Republic of Indonesia on "Cooperation on Reducing Greenhouse Gas Emissions from Deforestation and Forest Degradation".* Oslo, Norway.

Grieg-Gran, M., S. Bass, F. Booker, and M. Day. 2015. *The Role of Forests in a Green Economy Transformation in Africa.* Nairobi: United Nations Environment Programme.

Guillaud, D., 1994. Les douze ventres du sultan: la permanence des territoires à Jambi (Sumatra, Indonésie). *Géographie et Cultures*, 12, 109–130.

Gupta, J., 2012. Glocal forest and REDD+ governance: Win – win or lose – lose? *Current Opinion in Environmental Sustainability*, 4 (6), 620–627.

Hadiz V. R., 2001. Capitalism, oligarchic power, and the state in Indonesia. *Historical Materialism*, 8(1), 119–152.

Hagen, B., 1908. *Die Orang Kubu auf Sumatra.* Frankfurt am Main: Joseph Baer& CO.

Hall, D., 2013. *Land*. Cambridge, UK: Polity.

Hall, D., P. Hirsch, and T. M. Li. 2011. *Powers of Exclusion: Land Dilemmas in Southeast Asia*. Honolulu: University of Hawai'i Press.

Harvey, D., 2005. *A Brief History of Neoliberalism*. Oxford, UK: Oxford University Press.

Hauser-Schäublin, B., 2013. Introduction. The power of indigeneity: Reparation, readjustments and repositioning. *In*: B. Hauser-Schäublin, ed. *Adat and Indigeneity in Indonesia: Culture and Entitlements between Heteronomy and Self-Ascription*. Göttingen, Germany: Universitätsverlag Göttingen, 5–15.

Hauser-Schäublin, B., and S. Steinebach. 2014. *Harapan: A" no man' s land" turned into a contested agro-industrial Zone. Vol. No. 3, EFForts Discussion Paper Series*. Göttingen, Germany: University Göttingen.

Hein, J., 2013a. Climate change mitigation in emerging economies: The case of Indonesia: Hot air or leadership? *DIE Briefing Paper*, 8/2013.

Hein, J., 2013b. *Reducing Emissions from Deforestation and Forest Degradation (REDD+), transnational conservation and access to land in Jambi, Indonesia. Vol. No. 2, EFForts Discussion Paper Series*. Göttingen, Germany: University of Göttingen.

Hein, J., 2014. Politiken zur Reduktion von Emissionen aus Entwaldung und Schädigung von Wäldern (REDD+). *Peripherie: Zeitschrift für Politik und Ökonomie in der Dritten Welt*, (136), 508–511.

Hein, J., S. Adiwibowo, C. Dittrich, Rosyani, E. Soetarto, and H. Faust. 2016. Rescaling of access and property relations in a frontier landscape: Insights from Jambi, Indonesia. *The Professional Geographer*, 68 (3), 380–389.

Hein, J., and H. Faust. 2010. *Frontier Migration as response to environmental change*. Vol. No. 31 (July 2010), *STORMA Discussion Paper*. Göttingen, Germany: University of Göttingen.

Hein, J., and H. Faust. 2014. Conservation, REDD+ and the struggle for land in Jambi, Indonesia. *Pacific Geographies*, 41, 20–25.

Hein, J., H. Faust, Y. Kunz, and R. Mardiana. 2018a. The transnationalisation of competing state projects: Carbon offsetting and development in Sumatra's Coastal Peat Swamps. *Antipode*, 50, 953–975.

Hein, J., and H. Garrelts. 2014. Ambiguous involvement: Civil-society actors in forest carbon offsets. The case of the Climate Community and Biodiversity Standards (CCB). *In*: H. Garrelts and M. Dietz, eds. *Routledge Handbook of the Climate Change Movement*. New York: Routledge, 319–333.

Hein, J., A. Guarin, E. Frommé, and P. Pauw. 2018b. Deforestation and the Paris climate agreement: An assessment of REDD + in the national climate action plans. *Forest Policy and Economics*, 90, 7–11.

Hein, J., and Y. Kunz. 2018. Adapting in a carbon pool? Politicising climate change at Sumatra's oil palm frontier. *In*: S. Klepp and L. Chavez-Rodriguez, eds. *A Critical Approach to Climate Change Adaptation*. New York: Routledge, 151–169.

Hein, J., K. Meijer, and J. C. Rodriguez de Francisco. 2015. What is the potential for a climate, forest and community friendly REDD+ in Paris? *DIE Briefing Paper*, 3/2015.

Hidayat, R., 2012. *Membangkitkan Batang Terendam: Sejarah Asal Usl Kebudayaan dan Perjuangan Hak SAD Batin 9*. Jambi, Indonesia: Yayasan Setara Jambi.

Hiraldo, R., and T. Tanner. 2011. Forest voices: Competing narratives over REDD+. *IDS Bulletin*, 42 (3), 42–51.

Hirsch, J. (1995). *Der nationale Wettbewerbsstaat: Staat, Demokratie und Politik im globalen Kapitalismus*. Berlin and Amsterdam: Edition ID-Archiv.

Hirsch, J., and J. Kannankulam., 2011. The spaces of capital: The political form of capitalism and the internationalization of the state. *Antipode*, 43 (1), 12–37.

Heyde, J., M. C. Lukas, and M. Flitner. 2012. *Payments for Environmental Services in Indonesia: A Review of Watershed-related Schemes*. Artec-paper Nr. 186. Bremen, Germany: Artec – Forschungszentrum Nachhaltigkeit.

Heyvaert, V., 2017. The transnationalization of law: Rethinking law through transnational environmental regulation. *Transnational Environmental Law*, 6 (2), 205–236.

Hoey, B. A., 2003. Nationalism in Indonesia: Building imagined and intentional communities through transmigration. *Ethnology*, 42 (2), 109–126.

Hofman, B., and K. Kaiser. 2002. The making of the big bang and its aftermath, a political economy perspective. In: *A Conference Sponsored by the International Studies Program, Andrew Young School of Policy Studies*. Atlanta, Georgia: Georgia State University.

Holmgren, S., 2013. REDD+ in the making: Orders of knowledge in the climate – deforestation nexus. *Environmental Science & Policy*, 33, 369–377.

Horstmann, B., and J. Hein. 2017. *Aligning Climate Change Mitigation and Sustainable Development Under the UNFCCC: A Critical Assessment of the Clean Development Mechanism, the Green Climate Fund and REDD*. Bonn: Deutsches Institut für Entwicklungspolitik.

Houdret, A., I. Dombrowsky, and L. Horlemann. 2014. The institutionalization of River Basin Management as politics of scale – insights from Mongolia. *Journal of Hydrology*, 519, Part C, 2392–2404.

Houghton, R., B. Byers, and A. A. Nassikas. 2015. A role for tropical forests in stabilizing atmospheric CO 2. *Nature Climate Change*, 5 (12), 1022.

Howson, P., and S. Kindon. 2015. Analysing access to the local REDD+ benefits of Sungai Lamandau, Central Kalimantan, Indonesia. *Asia Pacific Viewpoint*, 56 (1), 96–110.

Hughes, D. M., 2005. Third nature: Making space and time in the Great Limpopo Conservation Area. *Cultural Anthropology*, 20 (2), 157–184.

HuMa., 2015. *Satu Tahun Pemerintahan Jokowi-JK: HuMa Meminta Pemerintah Hentikan Tarik Ulur dalam Penetapan Hutan Adat*. Available from: https://huma.or.id/uncategorized/satu-tahun-pemerintahan-jokowi-jk-huma-meminta-pemerintah-hentikan-tarik-ulur-dalam-penetapan-hutan-adat.html [Accessed 13 June 018].

ICAO (International Civil Aviation Organization). 2016. *Resolution A39-3: Consolidated statement of continuing ICAO policies and practices related to environmental protection – Global Market-based Measure (MBM) scheme* Available from: https://www.icao.int/environmental-protection/Documents/Resolution_A39_3.pdf [Accessed 21 August 2018].

Igoe, J., K. Neves and D. Brockington. 2010. A spectacular eco-tour around the historic bloc: Theorising the convergence of biodiversity conservation and capitalist expansion. *Antipode*, 42 (3), 486–512.

Indonesian REDD+ Task Force. 2012. *REDD+ National Strategy*. Jakarta, Indonesia: Indonesian REDD+ Task Force.

Indrarto, G. B., P. Murharjanti, J. Khatarina, I. Pulungan, F. Ivalerina, J. Rahman, M. N. Prana, I. A. P. Resosudarmo, and E. Muharrom. 2012. *The Context of REDD+ in Indonesia*. Bogor, Indonesia: Center for International Forestry Research (CIFOR).

Internationale Klimaschutzinitiative. 2015. *Harapan Rainforest – Pilothafte Restauration eines Degradierten Waldökosystems auf Sumatra*. Available from: www.international-climate-initiative.com/de/projekte/weltkarte-und-projektliste/details/272/ [Accessed 04 December 2015].

Internationale Klimaschutzinitiative. 2017. *Information for Applicants – Safeguards*. Available from: https://www.international-climate-initiative.com/en/project-funding/information-for-applicants/?iki_lang=en [Accessed 21 August 2018].

Ioris, A. A. R., 2014. *The Political Ecology of the State: The Basis and the Evolution of Environmental Statehood*. London, UK: Routledge.

Irfany, M. I., and S. Klasen. 2015. Inequality in emissions: Evidence from Indonesian household. *Environmental Economics and Policy Studies*, 1–25.

Jagger, P., K. Lawlor, M. Brockhaus, M. F. Gebara, D. J. Sonwa, and I. A. P. Resosudarmo. 2012. REDD+ safeguards in national policy discourse and pilot projects. *In*: A. Angelsen, M. Brockhaus, W. D. Sunderlin and L. Verchot, eds. *Analysing REDD+: Challenges and Choices*. Bogor, Indonesia: Center for International Forestry Research (CIFOR), 301–316.

JambiekspressNews. 2013. *19 SK Hutan Desa di Jambi Terancam Dicabut*. Available from: www.jambiekspresnews.com/berita-6661-19-sk-hutan-desa-di-jambi-terancam-dicabut. html [Accessed 04 December 2015].

JambiekspressNews. 2014. *Kajati: Tahura Ditangani Polda*. Available from: www.jambiekspres. co.id/berita-13758-kajati – tahura-ditangani-polda.html [Accessed 28 October 2015].

Jambipos Online. 2017. *SPI Tuding Oknum Pejabat Kanwil BPN Provinsi Jambi "Calo" Mafia Lahan di Jambi*. Available from: www.jambipos-online.com/2017/09/spi-tuding-oknum-pejabat-kanwil-bpn.html [Accessed 06 June 2018].

Jenkins, M., S. J. Scherr, and M. Inbar. 2004. Markets for biodiversity services: Potential roles and challenges. *Environment: Science and Policy for Sustainable Development*, 46 (6), 32–42.

Jessop, B., N. Brenner, and M. Jones. 2008. Theorizing sociospatial relations. *Environment and Planning. D, Society and space*, 26 (3), 389.

Jodoin, S., 2017. *Forest Preservation in a Changing Climate: REDD+ and Indigenous and Community Rights in Indonesia and Tanzania*. Cambridge, UK: Cambridge University Press.

Jodoin, S., and D. Mason-Case. 2016. What difference does CBDR make? A socio-legal analysis of the role of differentiation in the transnational legal process for REDD+. *Transnational Environmental Law*, 5 (2), 255–284.

Karriem, A., 2009. The rise and transformation of the Brazilian landless movement into a counter-hegemonic political actor: A Gramscian analysis. *Geoforum*, 40 (3), 316–325.

Kato, T., 1989. Different fields, similar locusts: Adat communities and the village law of 1979 in Indonesia. *Indonesia*, (47), 89–114.

Kawai, M., H. Scheyvens, H. Samejima, T. Fujisaki, and A. Setyarso. 2017. *Indonesia REDD+ Readiness: State of Play – March 2017*. Hayama: Institute for Global Environmental Strategies (IGES).

Kebschull, D., 1986. *Transmigration in Indonesia: An Empirical Analysis of Motivation, Expectations and Experiences*. Hamburg: Verlag Weltarchiv GmbH.

Kelly, A. B., 2011. Conservation practice as primitive accumulation. *Journal of Peasant Studies*, 38 (4), 683–701.

Kelly, P. F., 1997. Globalization, power and the politics of scale in the Philippines. *Geoforum*, 28 (2), 151–171.

Keohane, R. O., and J. S. Nye, Jr. 1998. Power and interdependence in the information age. *Foreign Affairs*, 81–94.

Kepala Desa Seponjen., 2013. *Rencana Pembangunan Jangka Menengah Desa (RPJM-DES) Seponjen*. Desa Seponjen, Jambi, Indonesia: Pemerintah Kabupaten Muaro Jambi.

Kesepakatan Terhadap Prasyarat Mediasi Antara PT. REKI Dengan Warga RT 11. 2012. (Unpublished document). Jambi, Indonesia.

Kijazi, M., 2015. Climate emergency, carbon capture and coercive conservation on Mt. Kilimanjaro. *In*: M. Leach and I. Scoones, eds. *Carbon Conflicts and Forest Landscapes in Africa*. London, UK and New York: Routledge, 58–78.

Koch, S., H. Faust, and J. Barkmann. 2008. Differences in power structures regarding access to natural resources at the village level in Central Sulawesi (Indonesia). *Austrian Journal of South-East Asian Studies*, 1 (2), 59–81.

Köhler, B., 2008. Die Materialität von Rescaling-Prozessen. Zum Verhälltnis von politics of scale und political ecology. *In*: M. Wissen, B. Röttger and S. Heeg, eds. *Politics of scale: Räume der Globalisierung und Perspektiven emanzipatorischer Politik*. Münster: Westfälisches Dampfboot, 208–223.

Kompas.com. 2016. *Jokowi: 12,7 Juta Hektar Lahan Akan Dibagikan kepada Masyarakat*. Available from: https://nasional.kompas.com/read/2016/12/30/12040991/jokowi.12.7.juta.hektar.lahan.hutan.akan.dibagikan.kepada.masyarakat [Accessed 13 June 2018].

Korhonen-Kurki, K., M. Brockhaus, E. Muharrom, S. Juhola, M. Moeliono, C. Maharani, B. Dwisatrio. 2017. Analyzing REDD+ as an experiment of transformative climate governance: Insights from Indonesia. *Environmental Science & Policy*, 73, 61–70.

Kosoy, N., and E. Corbera. 2010. Payments for ecosystem services as commodity fetishism. *Ecological Economics*, 69 (6), 1228–1236.

Kreditanstalt für Wiederaufbau. 2012. Financial cooperation with Indonesia (Unpublished document).

Kreditanstalt für Wiederaufbau. 2017. *Projektinformation: Lateinamerika – Waldschutz*. Available from: www.kfw-entwicklungsbank.de/PDF/Entwicklungsfinanzierung/Themen-NEU/Factsheets-COP23/2017_REM_Brasilien.pdf [Accessed 19 June 2018].

Krishna, V. V., U. Pascual, and M. Qaim., 2014. *Do emerging land markets promote forestland appropriation? Evidence from Indonesia. Vol. No. 7, EFForTS Discussion Paper Series*. Göttingen, Germany: University of Göttingen.

Kunz, Y., J. Hein, R. Mardiana, and H. Faust. 2016. Mimicry of the legal: Translating de jure land formalization processes into de facto local action – experiences from Jambi province, Sumatra, Indonesia. *Austrian Journal of South-East Asian Studies (ASEAS)*, 9 (1), 127–146.

Lang, C., 2010. *The Cancun Agreement on REDD: Four Questions and Four Answers*. Available from: www.redd-monitor.org/2010/12/18/the-cancun-agreement-on-redd-four-questions-and-four-answers/ [Accessed 03 December 2015].

Lang, C., 2012a. *Response from Germany's International Climate Initiative: The Mediation and Consultation Process at Harapan "Has been Rejected by the Groups Claiming Affiliation to SPI"*. Available from: www.redd-monitor.org/2012/12/21/response-from-germanys-international-climate-initiative/ [Accessed 29 May 2016].

Lang, C., 2012b. *Response from Harapan Rainforest Project: "The SPI Settlement Is Deep Inside Harapan, on a Scale Large Enough to Compromise the Ecological Integrity of the Forest"*. Available from: www.redd-monitor.org/2012/04/30/response-from-harapan-rainforest-project-the-spi-settlement-is-deep-inside-harapan-on-a-scale-large-enough-to-compromise-the-ecological-integrity-of-the-forest/ [Accessed 29 May 2016].

Lang, C., 2015. *Harapan Rainforest*. Available from: www.redd-monitor.org/?s=Harapan+Rainforest [Accessed 29 May 2016].

Langholz, J., J. Lassoie, and J. Schelhas. 2000. Incentives for biological conservation: Costa Rica's private wildlife refuge program. *Conservation Biology*, 14 (6), 1735–1743.

Larson, A. M., M. Brockhaus, W. D. Sunderlin, A. Duchelle, A. Babon, T. Dokken, T. T. Pham, I. Resosudarmo, G. Selaya, and A. Awono. 2013. Land tenure and REDD+: The good, the bad and the ugly. *Global Environmental Change*, 23 (3), 678–689.

Larson, A. M., and F. Soto., 2008. Decentralization of natural resource governance regimes. *Annual Review of Environment and Resources*, 33 (1), 213.

La Via Campesina. 2008. *Small Farmers Victims of Forest Carbon Trading*. Available from: www.viacampesina.org/en/index.php/actions-and-events-mainmenu-26/-climate-change-and-agrofuels-mainmenu-75/629-small-farmers-victims-of-forest-carbon-trading [Accessed 27 October 2015].

La Via Campesina. 2012. *Bangkok, UN Climate Negotiations Move Towards Burning the Planet.* Available from: www.viacampesina.org/en/index.php/actions-and-events-mainmenu-26/-climate-change-and-agrofuels-mainmenu-75/1296-bangkok-un-climate-negotiations-move-towards-burning-the-planet [Accessed 27 October 2015].

La Via Campesina. 2015. *Climate: Real Problem, False Solutions. No.3: REDD+.* Available from: https://viacampesina.org/en/climate-real-problem-false-solutions-no-3-redd/ [Accessed 20 August 2018].

Leach, M., R. Mearns, and I. Scoones. 1999. Environmental entitlements: Dynamics and institutions in community-based natural resource management. *World Development,* 27 (2), 225–247.

Leach, M., and Scoones, I. (Eds.). 2015. *Carbon Conflicts and Forest Landscapes in Africa.* New York: Routledge.

Lebel, L., P. Garden, and M. Imamura. 2005. The politics of scale, position, and place in the governance of water resources in the Mekong region. *Ecology and Society,* 10 (2), 18.

Lefebvre, H., 1976a. Reflections on the politics of space (translated by Enders, Michael J). *Antipode,* 8 (2), 30–37.

Lefebvre, H., 1976b. *The Survival of Capitalism.* London, UK: Allison & Busby.

Lefebvre, H., 1991. *The Production of Space.* Vol. 142. Oxford, UK: Blackwell Publishers Ltd.

Levang, P., and O. Sevin. 1990. *80 Years of Transmigration in Indonesia 1905–1985.* Jakarta: Departemen Transmigrasi and Institut Francais de Recherche Scientifique pour le Developpement en Cooperation (ORSTOM).

Li, T. M., 2001. Masyarakat adat, difference, and the limits of recognition in Indonesia's forest zone. *Modern Asian Studies,* 35 (3), 645–676.

Li, T. M., 2002. Local histories, global markets: Cocoa and class in upland Sulawesi. *Development and Change,* 33 (3), 415–437.

Li, T. M., 2005. *Transforming the Indonesian uplands.* Amsterdam, The Netherlands: Harwood Academic Publishers.

Li, T. M., 2008. Contested commodifications: Struggles over nature in a national park. *In*: J. Nevins and N. L. Peluso, eds. *Taking Southeast Asia to Market: Commodtities, Nature and People in the Neoliberal Age.* Ithaca: Cornell University Press, 124–139.

Locher-Scholten, E., 1994. Dutch expansion in the Indonesian archipelago around 1900 and the imperialism debate. *Journal of Southeast Asian Studies,* 25 (01), 91–111.

Locher-Scholten, E., 1996. *The establishment of colonial rule in Jambi: The dual strand of politics and economics.* Paper read at Colloquium, Historical foundations of a national economy in Indonesia, 1890s-1990s, at Amsterdam, New York.

Locher-Scholten, E., 2004. *Sumatran Sultanate and Colonial State: Jambi and the Rise of Dutch Imperialism, 1830–1907.* Ithaca, New York: SEAP Publications.

Lohmann, L., 2008. Carbon trading, climate justice and the production of ignorance: Ten examples. *Development,* 51 (3), 359–365.

Lubis, R., and I. Suryadiputra. 2004. Upaya pengelolaan terpadu hutan rawa gambut bekas terbakar diwilayah Berbak-Sembilang. *In*: S. Suyanto, U. Chokkalingam and P. Wibowo, eds. *Prosiding Semiloka "Kebakaran di Lahan Rawa/Gambut di Sumatera: Masalah dan Solusi".* Bogor, Indonesia: Center for International Forestry Research (CIFOR), 105–119.

Lukas, M. C., 2014. Eroding battlefields: Land degradation in Java reconsidered. *Geoforum,* 56, 87–100.

Lukes, S., 2005. *Power, Second Edition: A Radical View.* New York: Palgrave Macmillan.

Lund, C., 2006. Twilight institutions: Public authority and local politics in Africa. *Development and Change,* 37 (4), 685–705.

Lund, C., 2008. *Local Politics and the Dynamics of Property in Africa*. Cambridge, UK: Cambridge University Press.

Lund, C., 2016. Rule and rupture: State formation through the production of property and citizenship. *Development and Change*, 47 (6), 1199–1228.

Macpherson, C. B., 1978. *Property Mainstream and Critical Positions*. Oxford, UK: Basil Blackwell.

Makamah Konstitusi Republik Indonesia. 2012. *Putusan Nomor 35/PUU-X/2012*. Jakarta, Indonesia.

Manager of Royal Society for the Protection of Birds, R., 2012. Personal Communication, 08 July 2012.

Mann, G., 2009. Should political ecology be Marxist? A case for Gramsci's historical materialism. *Geoforum*, 40 (3), 335–344.

Mannell, J., 2014. Adopting, manipulating, transforming: Tactics used by gender practitioners in South African NGOs to translate international gender policies into local practice. *Health & Place*, 30, 4–12.

Marcus, G. E., 1995. Ethnography in/of the world system: The emergence of multi-sited ethnography. *Annual Review of Anthropology*, 25, 95–117.

Mardiana, R., 2014. Kehendak Merestorasi Ekosistem Tersandera di Pusaran Sengkarut Agraria: Konflik dan Perjuangan Kedaulatan Agraria di Wilayah Restorasi Ekosistem Hutan Harapan Provinsi Jambi. Vol. 14, Sayogyo Institute *Working Paper*. Bogor, Indonesia: Sayogyo Institute.

Mardiana, R., 2017. *Contesting Knowledge of Land Access Claims in Jambi, Indonesia*. Thesis (PhD). Georg-August-University Göttingen.

Marston, S. A., 2000. The social construction of scale. *Progress in Human Geography*, 24 (2), 219–242.

Marx, K. (1887). *Capital: A Critique of Political Economy. Volume I: The Process of Production of Capital*.

McAfee, K., 1999. Selling nature to save it? Biodiversity and green developmentalism. *Environment and Planning*, 17, 133–154.

McAfee, K., 2012a. The contradictory logic of global ecosystem services markets. *Development and Change*, 43 (1), 105–131.

McAfee, K., 2012b. Nature in the Market-World: Ecosystem services and inequality. *Development*, 55 (1), 25–33.

McCarthy, J. F., 2004. Changing to gray: Decentralization and the emergence of volatile socio-legal configurations in Central Kalimantan, Indonesia. *World Development*, 32 (7), 1199–1223.

McCarthy, J. F., 2005. Between adat and state: Institutional arrangements on Sumatra's forest frontier. *Human Ecology*, 33 (1), 57–82.

McCarthy, J. F., 2007. Shifting resource entitlements and governance reform during the agrarian transition in Sumatra, Indonesia. *Journal of Legal Pluralism*, 39 (55), 95–121.

McCarthy, J. F., 2012. Certifying in contested spaces: Private regulation in Indonesian forestry and palm oil. *Third World Quarterly*, 33 (10), 1871–1888.

McCarthy, J. F., C. Barr, I. A. P. Resosudarmo, and A. Dermawan. 2006. Origins and scope of Indonesia's decentralization laws. *In*: C. Barr, I. A. P. Resosudarmo and A. Dermawan, eds. *Decentralization of Forest Administration: Implications for Forest Sustainability, Economic Development and Community Livelihoods*. Bogor, Indonesia: Center for International Forestry Research (CIFOR), 31–57.

Mcculloch, L., 2010. Ulu Masen REDD Demonstration Project – The challenges of Tackling Market Policy and Governance Failures that Underlie Deforestation and Forest

Degradation. *Forest Conservation Team Occasional Papers*, Kamiyamaguchi, Japan: Institute for Global Environmental Strategies.

McGregor, A., 2010. Green and REDD? Towards a political ecology of deforestation in Aceh, Indonesia. *Human Geography*, 3 (2), 21–34.

Meadowcroft, J., 2002. Politics and scale: Some implications for environmental governance. *Landscape and Urban Planning*, 61 (2), 169–179.

Menteri Kehutanan. 1995. Keputusan Menteri Kehutanan Nomor: 592/Kpts-IV/1995 Tentang Pengesahan Rencana Karya Pengusahaan Hutan Yang Meliputi Seluruh Jangka Waktu Pengusahaan Hutan (Sementara) Atas Nama PT. Asialog Provinsi Jambi. In *No. 592/Kpts-IV/1995*. Jakarta, Indonesia.

Menteri Kehutanan. 2008a. Peraturan Menteri Kehutanan Tentang Hutan Desa. In *No. P. 49/Menhut-II/2008*. Jakarta, Indonesia.

Menteri Kehutanan. 2008b. Peraturan Menteri Kehutanan, Nomor: P. 68/Menhut-II/2008 tentang Penyelenggaraan Demonstration Activities Pengurangan Emisi Karbon Dari Deforestatsi Dan Degradasi Hutan. In *P.68/Menhut-II/2008*. Jakarta, Indonesia.

Menteri Kehutanan. 2009. Peraturan Menteri Kehutanan Republik Indonesia Nomor P. 36/Menhut-II/2009 Tentang Tata Cara Perizinan Usaha Pemanfaatan Penserapan Dan/Atau Penyimpanan Karbon Pada Hutan Produksi dan Hutan Lindung. In *P.36/Menhut-II/2009*. Jakarta, Indonesia.

Menteri Kehutanan. 2011. Peraturan Menteri Kehutanan Republik Indonesia Nomor: P55/Menhut-II/2011 Tentang Tata Cara Permohonan Izin Usaha Pemanfaatan Hasil Hutan Kayu Pada Hutan Tanaman Rakyat Dalam Hutan Tanaman. In *No. P. 55/Menhut-II/2011*. Jakarta, Indonesia.

Menteri Kehutanan. 2014. Peraturan Menteri Kehutanan Republik Indonesia Tentang Hutan Kemasyarakatan. In *No. P. 88/Menhut-II/2014*. Jakarta, Indonesia.

Menteri Pertanian. 2007. Peraturan Menteri Pertanian Tentang Pedoman Perizinan Usaha Perkebunan. In *No. 26/Permentan/OT.140/2/2007*. Jakarta, Indonesia.

Menteri Lingkungan Hidup dan Kehutanan. 2017. Peraturan Menteri Lingkungan Hidup dan Kehutanan Republik Indonesia Nomor: P.70/MENLHK/SETJEN/KUM.1/12/2017 Tentang Tata Cara Pelaksanaan Reducing Emissions from Deforestation and Forest Degradation, Role of Conservation, Sustainable Managment of Forest and Enhancement of Forest Carbon Stocks. In No. P.70 /MENLHK/SETJEN/KUM.1/12/2017. Jakarta, Indonesia.

Merry, S. E., 2000. Crossing boundaries: Ethnography in the twenty-first century. *PoLAR: Political and Legal Anthropology Review*, 23 (2), 127–133.

Mertz, O., 2009. Trends in shifting cultivation and the REDD mechanism. *Current Opinion in Environmental Sustainability*, 1 (2), 156–160.

Metzner, J., 1981. Palu (Sulawesi): Problematik der Landnutzung in einem klimatischen Trockental am Äquator. *Erdkunde*, 35 (1), 42–54.

Millennium Challenge Account. 2015. *Consolidation and Partnership for Progress Acceleration, Executive Summary MCA-Indonesia, Annual Report, April 2013 – December 2014*. Jakarta, Indonesia.

Ministry of Environment and Forestry. 2017. *Ecosystem Restoraton Progress*. Available from: www.iges.or.jp/files/research/natural-resource/PDF/20170615/E4_Happy.pdf [Accessed 01 June 2018].

Ministry of Forestry. 2007. *REDDI, Reducing Emissions from Deforestation and Forest Degradation in Indonesia*. Jakarta, Indonesia: Indonesia Forest Climate Alliance and Ministry of Forestry.

Ministry of Forestry. 2008. Ministry of Forestry Regulation No. P. 61/Menhut-II/2008 on Provisions and Pocedures for Issuing Ecosystem Restoration Forest Timber Utilisation

Permits for Natural Forests in Production Forests through Applications. In *No. P61/Menhut-II/2008.* Jakarta, Indonesia.

Ministry of Forestry. 2009. Ministry of Forestry Republic of Indonesia Decree Number P. 36/Menhut-II/2009 (unofficial translation) In *P.36/Menhut-II/2009.*

Moeliono, M., and A. Dermawan. 2006. The impacts of decentralization on tenure and Livelihoods. *In*: C. Barr, I. A. P. Resosudarmo, A. Dermawan and J. McCarthy, eds. *Decentralization of Forest Administration: Implications for Forest Sustainability, Economic Development and Community Livelihoods.* Bogor, Indonesia: Center for International Forestry Research (CIFOR), 108–120.

Mohsin, A., 2018. Coping with Indonesia's mudflow disaster. *In*: A. Sulfikar, ed. *The Sociotechnical Constitution of Resilience.* Singapore: Palgrave Macmillan, 117–145.

Moniaga, S., 1993. Toward community-based forestry and recognition of adat property rights in the outer Islands of Indonesia. *In*: J. Fox, ed. *Legal Frameworks for Forest Management in Asia: Case Studies of Community/ State Relations.* Occasional Paper No. 16, East-West Center, Honol Honolulu: East-West Center.

Moore, D. S., 1998. Subaltern struggles and the politics of place: Remapping resistance in Zimbabwe's eastern highlands. *Cultural Anthropology*, 13 (3), 344–381.

Muradian, R., M. Arsel, L. Pellegrini, F. Adaman, B. Aguilar, B. Agarwal, E. Corbera, D. Ezzine de Blas, J. Farley, G. Froger, E. Garcia-Frapolli, E. Gómez-Baggethun, J. Gowdy, N. Kosoy, J. F. Le Coq, P. Leroy, P. May, P. Méral, P. Mibielli, R. Norgaard, B. Ozkaynak, U. Pascual, W. Pengue, M. Perez, D. Pesche, R. Pirard, J. Ramos-Martin, L. Rival, F. Saenz, G. Van Hecken, A. Vatn, B. Vira, and K. Urama. 2013. Payments for ecosystem services and the fatal attraction of win-win solutions. *Conservation Letters*, 6 (4), 274–279.

Murdiyarso, D., S. Dewi, D. Lawrence, and F. Seymour. 2011. *Indonesia's Forest Moratorium: A Stepping Stone to Better Forest Governance?* Bogor, Indonesia: Center for International Forestry Research (CIFOR).

Murdiyarso, D., M. Brockhaus, W. D. Sunderlin and L. Verchot. 2012. Some lessons learned from the first generation of REDD+ activities. *Current Opinion in Environmental Sustainability*, 4(6), 678–685.

NABU. 2010. *Harapan – Hoffnung für Tiger & Co.* NABU-INFO 3/2010. Berlin: Naturschutzbund Deutschland Bundesverband.

NABU. 2014. *Projekt "Hoffnung", Bilanz nach fünf Jahren Wald- und Klimaschutz in Indonesien.* Available from: www.nabu.de/news/2014/03/16998.html [Accessed 06 June 2018].

The Nature Conservancy, without year. *Lore Lindu National Park, Building Partnerships to Protect Sulawesi's Unique Wildlife.* Park profile. Palu, Indonesia: The Nature Conservancy Indonesia Programme.

Naughton-Treves, L., and K. Wendland. 2014. Land tenure and tropical forest carbon management. *World Development*, 55, 1–6.

Neeff, T., L. Ashford, J. Calvert, C. Davey, J. Durbin, J. Ebeling, T. Herrera, T. Janson-Smith, B. Lazo, R. Mountain, S. O'Keeffe, S. Panfil, N. Thorburn, C. Tuite, M. Wheeland, and S. Young. 2009. *The Forest Carbon Offsetting Survey 2009*: Eco Securitas, Conservation International, The Climate, Community& Biodiversity Alliance, ClimateBiz.

Neumann, R. P., 2009. Political ecology: Theorizing scale. *Progress in Human Geography*, 33 (3), 398–406.

Nevins, J., and N. L. Peluso. 2008. Introduction: Commoditization in Southeast Asia. *In*: J. Nevins and N. L. Peluso, eds. *Taking Southeast Asia to Market.* Ithaca: Cornell University Press, 1–24.

Newell, P., and M. Paterson. 2010. *Climate Capitalism: Global Warming and the Transformation of the Global Economy*. New York: Cambridge University Press.

Nugraha, I., S. Sapariah, and A. A. Karokaro. 2017. *Bertemu AMAN, Presiden Kuatkan Komitmen Termasuk Segerakan Satgas Masyarakat Adat*. Available from: www.mongabay. co.id/2017/03/22/bertemu-aman-presiden-kuatkan-komitmen-termasuk-segerakan-satgas-masyarakat-adat/ [Accessed 13 June 2018].

Nuijten, M., 2005. Power in practice: A force field approach to natural resource management. *The Journal of Transdisciplinary Environmental Studies*, 4 (2), 1–14.

Nurhaniah. 2006. *Peran Kepala Desa Dalam Pendaftaran Hak Milik Atas Tanah Setelah Berlakunya* PP NO. 24 Tahun 1997 Di Kecamatan Tanah Grogot Kabupaten Pasir Kalimantan Timur, Program Pascasarjana Magister Kenotariatan Universitas Diponegoro, Semarang.

Nurjaya, I. N., 2005. Sejarah Hukum Pengelolaan Hutan di Indonesia. *Jurisprudence*, 2 (1), 35–55.

Ong, A., 2006. *Neoliberalism as Exception: Mutations in Citizenship and Sovereignty*. Durham, USA and London: Duke University Press.

Ortiz, S., 1984. Colonization in the Colombian Amazon. *In*: M. Schmink and C. H. Wood, eds. *Frontier Expansion in Amazonia*. Gainesville: University of Florida Press, 204–230.

Osborne, T. M., 2011. Carbon forestry and agrarian change: Access and land control in a Mexican rainforest. *Journal of Peasant Studies*, 38 (4), 859–883.

Paasi, A., 2003. Territory. *In*: J. Agnew and G. Toal, eds. *A Companion to Political Geoghraphy*. Malden, MA: Blackwell Publishing Ltd, 95–108.

Pagiola, S., 2011. *Using PES to Implement REDD, PES Learning Paper 2011–1*. Washington, DC: World Bank Publications.

Paoli, G. D., P. Gillespie, P. L. Wells, L. Hovani, A. Sileuw, N. Franklin, and J. Schweithelm. 2013. *Oil Palm in Indonesia: Governance, Decision Making, & Implications for Sustainable Development*. Jakarta, Indonesia: The Nature Conservancy.

Pasgaard, M., 2015. Lost in translation? How project actors shape REDD+ policy and outcomes in Cambodia. *Asia Pacific Viewpoint*, 56 (1), 111–127.

Pattberg, P., and J. Stripple. 2008. Beyond the public and private divide: Remapping transnational climate governance in the 21st century. *International Environmental Agreements: Politics, Law and Economics*, 8 (4), 367–388.

Peet, R., P. Robbins, and M. Watts. 2011. Global nature. *In*: R. Peet, P. Robbins and M. Watts, eds. *Global Political Ecology*. Milton Park: Routledge, 1–47.

Pellizzoni, L., 2011. Governing through disorder: Neoliberal environmental governance and social theory. *Global Environmental Change*, 21 (3), 795–803.

Peluso, N. L., 1992. *Rich Forest, Poor People*. Berkeley, CA: California University Press.

Peluso, N. L., 1995. Whose woods are these? Counter mapping forest territories in Kalimantan, Indonesia. *Antipode*, 27 (4), 383–406.

Peluso, N. L., S. Afiff, and N. F. Rachman. 2008. Claiming the grounds for reform: Agrarian and environmental movements in Indonesia. *Journal of Agrarian Change*, 8 (2–3), 377–407.

Peluso, N. L., and C. Lund. 2011. New frontiers of land control: Introduction. *Journal of Peasant Studies*, 38 (4), 667–681.

Pemerintah Kabupaten Batang Hari. 2010. *Rencana Strategi Penerapan Good Forestry Governance Kabupaten Batang Hari*. Muaro Bulian, Indonesia: Pemerintah Kabupaten Batang Hari.

Pemerintah Kabupaten Muaro Jambi. 2012. *Rencana Pembangunan JangkaPanjang Daerah (RPJPD) KabupatenMuaro Jambi, Tahun 2006–2025*. Sengeti, Jambi, Indonesia: BAPPEDA Muaro Jambi.

Pemerintah Kabupaten Batang Hari. 2012. *Sejarah Singkat Kabupaten Batang Hari.* Available from: www.batangharikab.go.id/bat/statis-7-sejarahberdirinyakabupatenbatanghari.html [Accessed 08 October 2015].

Pemerintah Provinsi Jambi. 2011. *Jambi Sebagai Provinsi Percontohan Untuk Mekanisme REDD+.* Jambi, Indonesia.

Pemerintah Provinsi Jambi. 2012. *Rencana Aksi Daerah Penurunan Emisi Gas Rumah Kaca (RAD GRK) Provinsi Jambi.* Jambi: Indonesia Pemerintah Provinsi Jambi.

Perbatakusuma, E. A., M. Ridwansyah, A. Akiefnawati, R. Widolo, W. Kurniawan, D. Primadona, M. Shakti, I. Andrian, D. Lindawati, and Alfiansyah. 2012. *Strategi dan Rencana Aksi Provinsi Jambi 2012–2030: Dokumen Risalah Eksekutif.* Jambi, Indonesia: Komisi Daerah REDD+ Jambi.

Perreault, T., 2003. Changing places: Transnational networks, ethnic politics, and community development in the Ecuadorian Amazon. *Political geography,* 22 (1), 61–88.

Pichler, M., 2014. *Politische Ökologie der Palmöl- und Agrartreibstoffproduktion in Südostasien.* Münster: Westfälisches Dampfboot.

Pichler, M., 2015. Legal dispossession: State strategies and selectivities in the expansion of Indonesian palm oil and agrofuel production. *Development and Change,* 46 (3), 508–533.

Plan Vivo. 2015. *Project Pipeline.* Available from: www.planvivo.org/project-network/project-pipeline/ [Accessed 04 December 2015].

Plan Vivo. 2017. History and Timeline. Available from: http://www.planvivo.org/about-plan-vivo/history-and-timeline/ [Accessed 23.08.2018].

Plan Vivo. 2018. *Durian Rambun, Indonesia.* Available from: www.planvivo.org/project-network/durian-rambun-indonesia/ [Accessed 01 June 2018].

Pokorny, B., J. Johnson, G. Medina, and L. Hoch. 2012. Market-based conservation of the Amazonian forests: Revisiting win – win expectations. *Geoforum,* 43 (3), 387–401.

Polsek Sungai Bahar. 2011. Data Profil Desa Dalam Wilayahm Sungai Bahar, Desa Tanjung Lebar. Bahar Selatan, Jambi, Indonesia.

Potter, L., 2012. New transmigration 'paradigm' in Indonesia: Examples from Kalimantan. *Asia Pacific Viewpoint,* 53 (3), 272–287.

Poudyal, M., B. Ramamonjisoa, N. Hockley, J. M. Gibbons, R. Mandimbiniaina, J. P. G. Jones. 2016. Can REDD+ social safeguards reach the 'right' people? Lessons from Madagascar. *Global Environmental Change,* 37, 31–42.

Poulantzas, N., 1978. *Staatstheorie, Politischer Überbau, Ideologie, Sozialistische Demokratie.* Hamburg: VSA-Verlag.

Presiden Republik Indonesia. 1960a. Peraturan Pemerintah Pengganti Undang-Undang Nomor 56 Tahun 1960. In *No 56/1960.* Jakarta, Indonesia.

Presiden Republik Indonesia. 1960b. Undang-Undang No.5 Tahun 1960 Tentang Peraturan Dasar Pokok-Pokok Agraria. In *No.5 1960.* Jakarta, Indonesia.

Presiden Republik Indonesia. 1967. Undang-Undang Nomor 5 Tahun 1967 Tentang Ketentuan-Ketentuan Pokok Kehutanan. In *No. 5/1967.* Jakarta, Indonesia.

Presiden Republik Indonesia. 1990. Peraturan Pemerintah Republik Indonesia Tentang Hak Pengusahaan Hutan Tanaman Industri. In *No. 7/1990.* Jakarta, Indonesia.

Presiden Republik Indonesia. 1997. Peraturan Pemerintah Republik Indonesia. In *No. 24/1997* Jakarta, Indonesia.

Presiden Republik Indonesia. 1999. Undang-Undang Republik Indonesia Nomor 41 Tahun 1999 Tentang Kehutanan. In *No 41/1999.* Jakarta, Indonesia.

Pundi Sumatera. 2014. *Project Identification Note (PIN) Community Forest Management Project in Jangkat Highland, JAMBI, Sumatra.* Merangin, Jambi, Indonesia: SSS-PUNDI.

Purnomo, H., D. Suyamto, L. Abdullah, and R. Irawati. 2012. REDD+ actor analysis and political mapping: An Indonesian case study. *International Forestry Review*, 14 (1), 74–89.

Purwanto, H., 2013. Local to global, how Serikat Petani Indonesia has accelerated the movement for Agrarian reform. *In*: La Via Campesina, ed. *La Via Campesina's Open Book: Celebrating 20 Years of Struggle and Hope*. Jakarta, Indonesia: La Via Campesina, 1–12.

Pye, O., 2012. Changing socio-natures in South-East Asia. *Austrian Journal of South-East Asian Studies*, 5 (2), 198.

Rachman, N. F., 2011. *The Resurgence of Land Reform Policy and Agrarian Movements in Indonesia*. Thesis (PhD). University of California, Berkeley, CA.

Rachman, N. F., 2013. *Undoing Categorical Inequality Masyarakat Adat, Agrarian Conflicts, and Struggle for Inclusive Citizenship in Indonesia, Paper Sajogyo Institute*. Bogor, Indonesia: Sajogyo Institute.

Rachman, N. F., and M. Siscawati. 2017. Forestry law, masyarakat adat and struggles for inclusive citizenship in Indonesia. *In*: C. Antons, ed. *Routledge Handbook of Asian Law*. New York: Routledge, 224–249.

Radjawali, I., O. Pye, M. Flitner. 2017. Recognition through reconnaissance? Using drones for counter-mapping in Indonesia. *The Journal of Peasant Studies*, 1–17.

The REDD Desk. 2015. *Financing Agreement RD3412*. Available from: http://theredddesk. org/countries/agreements/financing-agreement-rd3412 [Accessed 29 May 2016].

REDDplusid. 2016. *Pelajaran Resolusi Konflik Di Hutan Harapan*. Available from: www. reddplusid.org/eng/index.php/news/71-pelajaran-resolusi-konflik-di-hutan-harapan [Accessed 06 June 2018].

Reed, M. G., and S. Bruyneel. 2010. Rescaling environmental governance, rethinking the state: A three-dimensional review. *Progress in Human Geography*, 34 (5), 646–653.

REKI, PT Restorasi Ekosistem Indonesia. 2011a. Profil Masyarakat Desa di Sekitar Kawasan PT. Restorasi Ekosistem Indonesia – Desa Bungku, Durian Dangkal (Unpublished document). Bungku, Jambi, Indonesia.

REKI, PT Restorasi Ekosistem Indonesia. 2011b. *Surat Pernyataan*, Kepala Desa, ed. Bungku: PT REKI, Pemerintah Kabupaten Batanghari.

REKI, PT Restorasi Ekosistem Indonesia. 2013. Map of Encroachment 2005–2013 (Unpublished document). Bungku, Jambi, Indonesia.

REKI, PT Restorasi Ekosistem Indonesia. 2014. *Nature Tourism in the Harapan Rainforest. Harapan Rainforest Is a Paradise for Wildlife Adventurers*. Available from: http://harapanrainforest. org/harapan/subprogram/Nature%20Tourism%20in%20the%20Harapan%20Rainforest#. Vkxz_OLY5zk [Accessed 18 November 2015].

REKI, PT Restorasi Ekosistem Indonesia. 2015. *Komitmen Penghormatan Hak Asasi Manusi, Sosial dan Pelibatan Masyarakat PT. Restorasi Ekosistem Indonesia*. Available from: http:// hutanharapan.id/app/webroot/uploads/files/HARSCEC_%20Komitmen%20HAM_ Sosial_dan_Pelibatan_Masyarakat.pdf [Accessed 06 June 2018].

REKI, PT Restorasi Ekosistem Indonesia. 2017. *Kesepakatan Dengan Masyarakat Kunangan Jaya II*. Available from: http://hutanharapan.id/read/kesepakatan-dengan-masyarakat-kunangan-jaya-ii#.WxgLSSBpGF5 [Accessed 06 June 2018].

REKI, PT Restorasi Ekosistem Indonesia. 2018. *Agroforestri Adalah Masa Depan*. Available from: http://hutanharapan.id/read/agroforestri-adalah-masa-depan#.WxgNmCBpGF4 [Accessed 06 June 2018].

Republic of Indonesia. 1990. Undang-Undang Nomor 5/1990, Tentang Konservasi Sumber Daya Alam Hayati dan Ekosistemnya. Jakarta, Indonesia.

Republic of Indonesia. 2007. Lampiran, Peraturan Pemerintah Nomor 38/ 2007, Tentang Pembagian Urusan Pemerintahan Antara Pemerintah, Pemerintah Daerah Provinsi, dan Pemerintah Daerah Kabupaten/Kota. Jakarta, Indonesia.

Republic of Indonesia. 2011a. *Masterplan Acceleration and Expansion of Indonesia, Economic Development 2011–2025.* Jakarta, Indonesia: Coordinating Ministry for Economic Affairs.

Republic of Indonesia. 2011b. Presidential Regulation of The Republic of Indonesia No. 61 Year 2011 on The National Action Plan for Greenhouse Gas Emissions Reductions. In *No.61/2011.* Jakarta, Indonesia.

Republika Online. 2009. *150 KK Transmigrasi Segera Ditempatkan di Muarojambi.* Available from: www.republika.co.id/berita/breaking-news/nusantara/09/09/07/74563-150-kk-transmigrasi-segera-ditempatkan-di-muarojambi [Accessed 26 June 2018].

Resosudarmo, I. A. P., 2004. Closer to people and trees: Will decentralisation work for the people and the forests of Indonesia? *The European Journal of Development Research*, 16 (1), 110–132.

Resosudarmo, I. A. P., S. Atmadja, A. D. Ekaputri, D. Y. Intarini, Y. Indriatmoko, and P. Astri. 2014. Does tenure security lead to REDD+ project effectiveness? Reflections from five emerging sites in Indonesia. *World Development*, 55, 68–83.

Resosudarmo, I. A. P., C. Barr, A. Dermawan, and J. McCarthy. 2006. Fiscal balancing and the redistribution of forest revenues. *In*: C. Barr, I. A. P. Resosudarmo, A. Dermawan and J. McCarthy, eds. *Decentralization of Forest Administration in Indonesia: Implications for Forest Sustainability, Economic Development and Community Livelihoods.* Bogor, Indonesia: Center for International Forest Research (CIFOR), 58–86.

Rettet den Regenwald. 2014. *Palmölfirma PT Asiatic Persada 3 Jahrzehnte Landraub, Vertreibung, Menschenrechtsverletzungen, Gewalt und Mord. Eine Chronologie.* Available from: www.regenwald.org/files/de/Chronic-Asiatic-Persada.pdf [Accessed 27 October 2015].

Rhee, S., 2009. The cultural politics of collaboration to control and access forest resources in Malinau, East Kalimantan. *In*: M. Moeliono, E. Wollenberg and G. Limberg, eds. *The Decentralization of Forest Governance – Politics, Economics and the Fight for Control of Forests in Indonesian Borneo.* London, UK: Earthscan Publishers, 43–59.

Ribot, J. C., and N. L. Peluso. 2003. A theory of access*. *Rural sociology*, 68 (2), 153–181.

Rice, R., 2002. *Conservation Concessions – Concept Description.* Washington, DC: Center for Applied Biodiversity Science at Conservation International.

Robinson, W. I., 2001 Social theory and globalization: The rise of a transnational state. *Theory and Society*, 30 (2), 157–200.

Robin Wood. 2011. Statement of ROBIN WOOD to the report of TÜV Rheinland. Available from: https://www.robinwood.de/uploads/media/Statement_Robin_Wood [Accessed 21 August 2018].

Robin Wood. 2014. Palmöl tötet. Available from: https://www.robinwood.de/fileadmin/Redaktion/Dokumente/Magazin/2014-2/121-44-47-imu3.pdf [Accessed 21 August 2018].

Rodríguez de Francisco, J. C., 2013. *PES, peasants and power in Andean watersheds: Power relations and payment for environmental services in Colombia and Ecuador.* Thesis (PhD). Wageningen School of Social Science, Wageningen University, Wageningen, The Netherlands.

Rodriguez de Francisco, J. C., and R. Boelens. 2014. Payment for environmental services and power in the Chamachán watershed, Ecuador. *Human Organization*, 73 (4), 351–362.

Rodríguez de Francisco, J. C., J. Budds, and R. Boelens. 2013. Payment for environmental services and unequal resource control in Pimampiro, Ecuador. *Society & Natural Resources*, 26 (10), 1217–1233.

Roth, D., 2009. Property and authority in a migrant society: Balinese irrigators in Sulawesi, Indonesia. *Development and Change*, 40 (1), 195–217.

RSPB, Royal Society for the Protection of Birds. 2015. *Annual Review, 2014–2015*. Available from: www.rspb.org.uk/about/run/annualreview/2015/index.html [Accessed 04 December 2015].

Sammukri, M. H., 2013. Mobilities of Indigeneity: Intermediary NGOs and Indigenous peoples in Indonesia. *In:* B. Hauser-Schäublin, ed. *Adat and Indigeneity in Indonesia: Culture and Entitlements Between Heteronomy and Self-Ascription*. Göttingen, Germany: Universitätsverlag Göttingen, 115–131.

Sargent, S., 2015. The contested meaning of free, prior and informed consent in international financial law and Indigenous rights. *In:* V. Vadi and D. B. Witte, eds. *Culture and International Economic Law*. London, UK and New York: Routledge, 87–103.

Sassen, S., 2008. *Territory, Authority, Rights: From Medieval to Global Assemblages*. Princeton: Princeton University Press.

Schielmann, S., M. Degawan, E. Falley-Rothkopf, B. Henneberger, K. Mantzel, and U. Nolte. 2013. *Waldschutzvorhaben im Rahmen der Klimapolitik und die Rechte indigener Völker*. Cologne, Germany: Institut für Ökologie und Aktions-Ethnologie e.V.

Schwarze, S., M. Euler, M. Gatto, J. Hein, E. Hettig, A. M. Holtkamp, L. Izhar, Y. Kunz, J. Lay, and J. Merten. 2015. *Rubber vs. oil palm: An analysis of factors influencing smallholders' crop choice in Jambi, Indonesia. Vol. No. 11, EFForTS Discussion Paper Series*. Göttingen, Germany: University of Göttingen.

Scott, J. C., 1985. *Weapons of the Weak: Everyday forms of Peasant Resistance*. New Haven, CT: Yale University Press.

Scott, J. C., 1989. Everyday forms of resistance. *Copenhagen Journal of Asian Studies*, 4, 33–62.

Serikat Petani Indonesia. 2012. *Dana Pengalihan Hutang Untuk Mengusir Petani*. Available from: www.spi.or.id/dana-pengalihan-hutang-untuk-mengusir-petani/ [Accessed 27 November 2015].

Sevin, O., and D. Benoît. 1993. Techniques d'encadrement et Terres-Neuves: les enseignements du delta du Batang Hari (Jambi-Indonésie). *Géographie et Cultures*, (7), 93–112.

Sikor, T., and C. Lund. 2010. Access and property: A question of power and authority. *In:* T. Sikor and C. Lund, eds. *The Politics of Possession: Property, Authority, and Access to Natural Resources*. West Sussex, UK: Wiley-Blackwell, 1–22.

Silalahi, M., and D. Erwin. 2013. *Collaborative conflict management on ecosystem restoration (ER) area: Lessons learnt from Harapan Rainforest (HRF) Jambi*. Paper read at International Conference of Indonesia Forestry Researchers, 2nd INAFOR, at Menara Peninsula, Jakarta, Indonesia.

Singapore Airlines. 2010. *Singapore Airlines commits to rainforest preservation*. Available from: https://www.singaporeair.com/en_UK/sg/media-centre/press-release/article/?q=en_UK/2010/July-September/20Aug2010-1102 [Accessed 21 August 2018].

Singapore Airlines. 2015. *Our Commitment to the Environment*. Available from: www.singaporeair.com/en_UK/about-us/sia-history/sia-environment/ [Accessed 27 October 2015].

Smith, J., K. Obidzinski, S. Subarudi, and I. Suramenggala. 2003. Illegal logging, collusive corruption and fragmented governments in Kalimantan, Indonesia. *International Forestry Review*, 5 (3), 293–302.

Smith, N., 1992. Contours of a spatialized politics: Homeless vehicles and the production of geographical scale. *Social Text*, (33), 55–81.

Smith, N., 2008. *Uneven Development: Nature, Capital, and the Production of Space*. Athens: University of Georgia Press.

Soja, E. W., 1980. The socio-spatial dialectic. *Annals of the Association of American Geographers*, 70 (2), 207–225.

Soja, E. W., 1989. *Postmodern Geographies: The Reassertion of Space in Critical Social Theory.* London, UK: Verso.

Soja, E. W., 1996. *Thirdspace: Journeys to Los Angeles and other Real-and-Imagined Places.* Oxford, UK: Basil Blackwell.

Spiller, I., and L. Fuhr. 2010. *Wo steht die internationale Klimapolitik nach Cancun? Heinrich Böll Stiftung.* Available from: www.boell.de/de/navigation/klima-energie-analyse-klimagipfel-cancun-10814.html [Accessed 03 December 2015].

Steinebach, S., 2013a. *Der Regenwald ist unser Haus: die Orang Rimba auf Sumatra zwischen Autonomie und Fremdbestimmung.* Göttingen, Germany: Universitätsverlag Göttingen.

Steinebach, S., 2013b. "Today we Occupy the Plantation – Tomorrow Jakarta": Indigeneity, land and oil palm plantations in Jambi. *In*: B. Hauser-Schäublin, ed. *Adat and Indigeneity in Indonesia – Culture and Entitlements between Heteronomy and Self-Ascription.* Göttingen, Germany Universitätsverlag Göttingen, 63–70.

Steinebach, S., forthcoming. *Reproduction of Ethninized Territories.*

Stephan, B., and R. Lane., 2014. Zombie markets or zombie analyses? Revivifying the politics of carbon markets. *In*: B. Stephan, R. Lane, eds. *The Politics of Carbon Markets.* London, UK: Routledge, 15–38.

Stern, N., 2007. *Stern Review: The Economics of Climate Change.* Cambridge, UK: Cambridge University Press.

STN. 2018. *Tentang Serikat Tani Nasional.* Available from: www.stn.or.id/tentang-stn/?i=1 [Accessed 06 June 2018].

Suprapto, Y., 2016. *Nasib Hutan Harapan dengan Segudang Masalah Lahan (Bagian 2).* Available from: www.mongabay.co.id/2016/10/03/nasib-hutan-harapan-dengan-segudang-masalah-lahan/ [Accessed 06 June 2018].

Swyngedouw, E., 2000. Authoritarian governance, power, and the politics of rescaling. *Environment and Planning D*, 18 (1), 63–76.

Swyngedouw, E., 2004. Globalisation or 'glocalisation'? Networks, territories and rescaling. *Cambridge Review of International Affairs*, 17 (1), 25–48.

Swyngedouw, E., 2010. *Place, Nature and the Question of Scale: Interrogating the Production of Nature.* Berlin: Berlin-Brandenburgische Akademie der Wissenschaften.

Szczepanski, K., 2002. Land policy and adat law in Indonesia's forests. *Pacific Rim Law& Policy Journal*, 11, 231–255.

Tambunan, I., 2015. 3.200 Perambah Liar di TN Kerinci Seblat Tuntut Akomodasi Hak Kependudukan. *Kompas.*

TFCA-Sumatera, Tropical Forest Conservation Action-Sumatra. 2014. *Selamat Datang di TFCA Sumatera.* Available from: www.tfcasumatera.org/2014/?lang=en [Accessed 04 December 2015].

Thorburn, C. C., 2004. The plot thickens: Land administration and policy in post-new order Indonesia. *Asia Pacific Viewpoint*, 45 (1), 33–49.

Tidemann, J., 1938. *Djambi.* Amsterdam, The Netherlands: Koloniaal Instituut.

To, P., W. Dressler, and S. Mahanty. 2017. REDD+ for Red Books? Negotiating rights to land and livelihoods through carbon governance in the Central Highlands of Vietnam. *Geoforum*, 81, 163–173.

Toumbourou, T., 2015. *Fires Spark Hope for Improved Law Enforcement.* Available from: http://indonesiaatmelbourne.unimelb.edu.au/fires-spark-hope-for-improved-law-enforcement/ [Accessed 11 November 2015].

Towers, G., 2000. Applying the political geography of scale: Grassroots strategies and environmental justice. *The Professional Geographer*, 52 (1), 23–36.

Tsing, A. L., 2005. *Friction: An Ethnography of Global Connection*. Princeton: Princeton University Press.

Tuong, V., 2009. Indonesia's agrarian movement: Anti-capitalism at a crossroads. *In*: D. Caouette and S. Turner, eds. *Agrarian Angst and Rural Resistance in Contemporary Southeast Asia*. Abingdon, UK: Routledge, 180–205.

Turner, S., and D. Caouette. 2009. Shifting fields of rural resistance in Southeast Asia. *In*: D. Caouette and S. Turner, eds. *Agrarian Angst and Rural Resistance in Contemporary Southeast Asia*. Abingdon, UK: Routledge, 1–24.

Ufen, A., 2002. *Herrschaftsfiguration und Demokratisierung in Indonesien (1965–2000)*. Hamburg, Germany: Institut für Asienkunde.

UNFCCC, United Nations Framework Convention on Climate Change. 2007. *Decision 2/CP.13 Reducing emissions from deforestation in developing countries: Approaches to stimulate action*. Bonn, Germany.

UNFCCC, United Nations Framework Convention on Climate Change. 2010. Decision 1/CP.16 The Cancun Agreements: Outcome of the work of the Ad Hoc Working Group on Long-term Cooperative Action under the Convention. Bonn, Germany.

UNFCCC, United Nations Framework Convention on Climate Change. 2013. *Warsaw Framework for REDD-plus*. Available from: http://unfccc.int/land_use_and_climate_change/redd/items/8180.php [Accessed 03 December 2015].

UN REDD. 2012. *Integrating Gender into REDD+ Safeguards Implementation in Indonesia*.

UN REDD. 2013. *Guidelines on Free, Prior and Informed Consent*. Geneva, Switzerland.

Usman, M., 2012. *Pengusiran Petani REKI Dimulai, 3 Rumah Dirobohkan*. Available from: www.metrojambi.com/v1/daerah/13120-pengusiran-petani-reki-dimulai-3-rumah-dirobohkan.html [Accessed 18 September 2015].

van Meijl, T., and F. von Benda-Beckmann. 1999. Introduction. *In*: T. van Meijl and F. von Benda-Beckmann, eds. *Property Rights and Economic Development Land and Natural Resources in Southeast Asia and Oceania*. London, UK: Kegan Paul International, 1–14.

VCS, Verified Carbon Standard. 2015. *What is the voluntary carbon market* [online]. Available from: http://www.v-c-s.org/faqs/what-voluntary-carbon-market. [Accessed 11 December 2015].

Virgilio, N. R., S. Marshall, O. Zerbock, and C. Holmes. 2010. *Reducing Emissions from Deforestation and Degradation (REDD): A Casebook of On-the-Ground Experience. 2010*. Arlington, VA: The Nature Conservancy, Conservation International and Wildlife Conservation Society.

Visseren-Hamakers, I. J., C. McDermott, M. J. Vijge, and B. Cashore. 2012. Trade-offs, co-benefits and safeguards: Current debates on the breadth of REDD+. *Current Opinion in Environmental Sustainability*, 4 (6), 646–653.

von Benda-Beckmann, F., and K. von Benda-Beckmann. 1999. A functional analysis of property rights, with special reference to Indonesia. *In*: T. van Meijl and F. von Benda-Beckmann, eds. *Property Rights and Economic Development Land and Natural Resources in Sotheast Asia and Oceania*. London, UK: Kegan Paul International, 15–56.

von Benda-Beckmann, F., K. von Benda-Beckmann, and A. Griffiths. 2005. Mobile people, mobile law: An introduction. *In*: F. Benda-Beckmann, K. Benda-Beckmann and A. Griffiths, eds. *Mobile People, Mobile Law: Expanding Legal Relations in a Contracting World*. Aldershot, UK: Ashgate, 1–25.

von Benda-Beckmann, F., K. von Benda-Beckmann, and M. Wiber. 2009. The properties of property. *In*: F. von Benda-Beckmann, K. von Benda-Beckmann and M. Wiber, eds. Changing *properties of property*. New York and Oxford, UK: Berghahn Books, 1–39.

von Benda-Beckmann, K., 1981. Forum shopping and shopping forums: Dispute processing in a Minangkabau village in West Sumatra. *The Journal of Legal Pluralism and Unofficial Law*, 13 (19), 117–159.

WALHI Jambi. 2009. *Alihfungsi Tahura Sungai Aur Jadi Trans dan kebun sawit.* Available from: http://walhi-jambi.blogspot.de/2009/02/alihfungsi-tahura-sungai-aur-jadi-trans.html [Accessed 20 January 2014].

WALHI Jambi. 2016. *Walhi Sayangkan Konsorsium NGO di PT ABT.* Available from: www.walhi-jambi.com/2016/02/walhi-sayangkan-konsorsium-ngo-di-pt-abt.html [Accessed 01 June 2018].

Walsh, T. A., Y. H. Asmui, and A. B. Utomo. 2012a. 1.5 Ecosystem restoration in Indonesia's production forests: Towards financial feasibility. *Etern News*, (54), 35–41.

Walsh, T. A., Y. H. Asmui, and A. Utomo. 2012b. *Supporting Ecosystem Restoration Concessions in Indonesia's Production Forests: A Review of the Licensing Framework 2004–2012*: Burung Indonesia-Climate and Land Use Alliance.

Wardah, N., 2013. *Harapan Rainforest: Cultivating or Ignoring Hopes.* Thesis (MsC). Rural Development Sociology Group, Wageningen University, Wagenigen, The Netherlands.

Warren, C., 1990. *The bureaucratisation of local government in Indonesia.* Vol. 66, *Working Paper.* Clayton: Australia The Centre of Southeast Asian Studies Monash University.

Weber, M., 1993. *Basic Concepts in Sociology.* New York: Citadel Press.

Wells, M., S. Guggenheim, A. Khan, W. Wardojo, and P. Jepson, P. 1999. *Investing in Biodiversity: A Review of Indonesia's Integrated Conservation and Development Projects.* Washington, DC: World Bank Publications.

While, A., A. E. Jonas, and D. Gibbs. 2010. From sustainable development to carbon control: Eco-state restructuring and the politics of urban and regional development. *Transactions of the Institute of British Geographers*, 35 (1), 76–93.

Wibisono, A. R., 2012. Pelaksanaan Pendaftaran Tanah Pertama Kali Secara Sporadik Menjadi Sertifkat Hak Milik Berdasarkan Surat Segel Fakultas Hukum Unversitas Brawijaya Malang. Indonesia.

Wibowo, A., and L. Giessen. 2015. Absolute and relative power gains among state agencies in forest-related land use politics: The Ministry of Forestry and its competitors in the REDD+ programme and the one map policy in Indonesia. *Land Use Policy*, 49, 131–141.

Wimmer, A., 2008. The making and unmaking of ethnic boundaries: A multilevel process theory. *American Journal of Sociology*, 113 (4), 970–1022.

Wirasapeotra, K., and H. Octavian. 2012. *Ringkasan Hasil Studi Identifikasi dan Analisa Konflik Para Pihak di Proyek Restorasi Ekosistem.* Pekanbaru, Riau, Indonesia: Scale-Up.

Wissen, M., 2008. Zur räumlichen Dimensionierung sozialer Prozesse. Die Scale-Debatte in der angloamerikanischen Radical Geography – eine Einleitung. *In*: M. Wissen, B. Röttger and S. Heeg, eds. *Politics of Scale: Räume der Globalisierung und Perspektiven emanzipatorischer Politik.* Münser: Westfälisches Dampfboot, 8–32.

Wissen, M., 2011. *Gesellschaftliche Naturverhältnisse in der Internationalisierung des Staates: Konflikte um die Räumlichkeit staatlicher Politik und die Kontrolle natürlicher Ressourcen.* Münster: Westfälisches Dampfboot.

Wolman, A., 2004. Review of conservation payment initiatives in Latin America: Conservation concessions, conservation incentive agreements and permit retirement schemes. *William and Mary Environmental Law and Policy Review*, 28, 859–981.

The World Bank. 1979. *Indonesia Transmigration II Staff Appraisal Report*: Regional Projects Department East Asia and Pacific Regional Office.

Wynberg, R., and M. Hauck. 2014. People, power, and the Coast: A conceptual framework for understanding and implementing benefit sharing. *Ecology and Society*, 19 (1).

Zainuddin. 2013. *PT Asiatic Persada dan Permasalahannya.* Jambi, Indonesia: KKI Warsi and Wilmar.

Zelli, F., D. Erler, S. Frank, J. Hein, H. Hotz, and A. M. Santa Cruz Melgarejo. 2014. *Reducing Emissions from Deforestation and Degradation in Peru – A challenge to social inclusion and multi-level governance*: *DIE Studies*. Bonn: DIE.

Zimmerer, K. S., 2000. The reworking of conservation geographies: Nonequilibrium landscapes and nature-society hybrids. *Annals of the Association of American Geographers*, 90 (2), 356–369.

Zimmerer, K. S., 2006. Cultural ecology: At the interface with political ecology-the new geographies of environmental conservation and globalization. *Progress in Human Geography*, 30 (1), 63.

ZGF, Zoologische Gesellschaft Frankfurt. 2016. *Waldmenschen im Babyglück*. Available from: https://fzs.org/de/aktuelles/waldmenschen-im-babyglueck/ [Accessed 01 June 2018].

ZSL, The Zoological Society of London. 2008. *Darwin Initiative Application for Grant for Round 16: Stage 2*. London, UK: Darwin Initiative and Department for Environment, Food and Rural Affairs.

ZSL, The Zoological Society of London. 2015. *About the Zoological Society of London*. Available from: www.zsl.org/about-us [Accessed 04 December 2015].

Zulu, L. C., 2009. Politics of scale and community-based forest management in southern Malawi. *Geoforum*, 40 (4), 686–699.

Index

For Product Safety Concerns and Information please contact our EU
representative GPSR@taylorandfrancis.com
Taylor & Francis Verlag GmbH, Kaufingerstraße 24, 80331 München, Germany

www.ingramcontent.com/pod-product-compliance
Lightning Source LLC
Chambersburg PA
CBHW071420180526
45170CB00001B/163